Concurrent Learning and Information Processing

Concurrent Learning and Information Processing

A Neuro-Computing System that Learns *During* Monitoring, Forecasting, and Control

ROBERT J. JANNARONE

President and Founder
Rapid Clip Neural Systems, Inc.
Atlanta, GA

CHAPMAN & HALL

ITP® INTERNATIONAL THOMSON PUBLISHING

New York • Albany • Bonn • Boston • Cincinnati • Detroit • London
Madrid • Melbourne • Mexico City • Pacific Grove • Paris • San Francisco
Singapore • Tokyo • Toronto • Washington

Cover design: Curtis Tow Graphics
e-ISBN-13: 978-1-4613-0431-9
DOI: 10.1007/ 978-1-4613-0431-9

Softcover reprint of the hardcover 1st edition 1997

Chapman & Hall
115 Fifth Avenue
New York, NY 10003

Chapman & Hall
2-6 Boundary Row
London SE1 8HN
England

Thomas Nelson Australia
102 Dodds Street
South Melbourne, 3205
Victoria, Australia

Chapman & Hall GmbH
Postfach 100 263
D-69442 Weinheim
Germany

International Thomson Editores
Campos Eliseos 385, Piso 7
Col. Polanco
11560 Mexico D.F
Mexico

International Thomson Publishing–Japan
Hirakawacho-cho Kyowa Building, 3F
1-2-1 Hirakawacho-cho
Chiyoda-ku, 102 Tokyo
Japan

International Thomson Publishing Asia
221 Henderson Road #05-10
Henderson Building
Singapore 0315

1 2 3 4 5 6 7 8 9 10 XXX 01 00 99 98 97

Library of Congress Cataloging-in-Publication Data

Jannarone, Robert J.
Concurrent learning and information processing : a neuro-computing system that learns during monitoring, forecasting, and control / Robert J. Jannarone.
p. cm.
Includes bibliographical references and index.
ISBN-13: 978-1-4613-8049-8 (hb : alk. paper)
1. Neural networks (Computer Science) 2. Parallel processing (Electronic computers) 3. Machine learning. I. Title.
QA76.87.J36 1997
003.'3632--dc21 97-5551
CIP

British Library Cataloguing in Publication Data available

To order this or any other Chapman & Hall book, please contact **International Thomson Publishing, 7625 Empire Drive, Florence, KY 41042.** Phone: (606) 525-6600 or 1-800-842-3636. Fax: (606) 525-7778. e-mail: order@chaphall.com.

For a complete listing of Chapman & Hall titles, send your request to **Chapman & Hall, Dept. BC, 115 Fifth Avenue, New York, NY 10003.**

To those who support progress when the going gets tough.

Contents

Dedication v
Preface ix

PART 1. RAPID LEARNING BENEFITS 2

Chapter 1. Rapid Learning Applications 3
Chapter 2. Contrasts With Alternative Methods 29

PART 2. RAPID LEARNING FEATURES 61

Chapter 3. The Rapid Learning Neuro-Computing System 63
Chapter 4. Rapid Learning Models 99

PART 3. RAPID LEARNING FOUNDATIONS 131

Chapter 5. Basic Concurrent Learning and Prediction 133
Chapter 6. Extended Concurrent Learning and Prediction 173

PART 4. OPERATIONAL DETAILS 211

Chapter 7. Monitoring and Forecasting Details 213
Chapter 8. Control and Refinement Details 229

Glossary 249
Index 259

Preface

GOALS

The primary goal of this book is to present a concurrent learning and information processing system for adaptive monitoring, forecasting, and control applications. A second goal is to generate related interest among information scientists.

This book is *not* a comprehensive survey of neural networks, machine learning models, psychological learning theory, or statistical methods. This book does describe ties between these fields and the system being presented, but presenting one system for rapid learning applications is its primary aim.

Above all, this book is not about a collection of abstract, theoretical models. It *is* about one practical solution to the main problem of the information age: "*So much to learn . . . so little time*."

FORMAT

To meet its main goal, this book presents examples of *working* solutions to *practical* information processing problems. The same examples are used to unify functional descriptions, software introductions, and technical treatments. The examples are also used to compare the methods in this book with alternative methods. In addition, the book introduces operational *Rapid Learner*™ software for solving concurrent learning and information processing problems.

Throughout this book, concurrent learning and information processing methods, along with related neuro-computing systems, are abbreviated as rapid learning methods and systems.

To meet its second goal, the book outlines rapid learning foundations from statistical, psychological, neuro-biological, and computing perspectives. The book also provides sufficient detail for scientists to understand concurrent learning concepts at a technical level. In addition, all chapters include Future Directions sections that are intended as food for scientific thought.

In keeping with its problem-solving focus, this book proceeds from rapid learning benefits (Part 1) to features (Part 2) to foundations (Part 3) to details (Part 4). Part 1, Rapid Learning Benefits, includes surveys of on-line learning applications (Chapter 1), as well as contrasts with alternative methods (Chapter 2). Part 2, Rapid Learning Features, includes an overview of the concurrent learning and information processing computer system (Chapter 3), especially *Rapid Learner*™ software. The software is covered in considerable detail because all examples in this book have been obtained from it and simulation versions of it are being prepared for use on the Internet (software information may be obtained *via* the Internet at http//www.rapidlearner.com). Part 2 also includes a description of rapid learning models (Chapter 4), including the rapid learning neuro-computing model and closely related scientific models.

Part 3 and Part 4 explain concurrent learning methods in greater detail. Part 3, Rapid Learning Foundations, describes concurrent learning methods in terms of linear statistical models and methods (Chapter 5), as well as categorical, non-linear, and other extensions (Chapter 6). Part 4, Operational Details, includes detailed explanations of the main information processing procedures that operate concurrently with learning: monitoring for unusual measurement values along with forecasting future measurements (Chapter 7); and controlling ongoing processes along with refining performance models (Chapter 8). To make descriptions precise, sections including detailed mathematical formulas have been included in Parts 3 and 4.

A Glossary at the end of this book includes a brief definition of each key technical term. Any term that is underlined in the book is defined in the Glossary.

USE

Readers who wish to use rapid learning methods should pay special attention to the practical examples in this book. Readers are especially encouraged to cover Chapter 1 through Chapter 3, which focus on practical solutions to information processing problems. Some of the detailed explanations in later chapters contain technical material that is not essential for practical use of rapid learning methods. For example, readers interested only in rapid learning applications may skip the optional mathematical formulations in Parts 3 and 4.

Information scientists who want to understand concurrent learning theory should pay special attention to the model descriptions in Chapter 4, as well as the detailed mathematical formulations in Parts 3 and 4. Extending the perspectives that are outlined in Chapter 4 into related fields will also be a useful exercise for them. Also, the Future Directions sections in the Conclusions to each chapter have been written specifically for information scientists. Identifying and actively pursuing future directions that interest them personally is especially recommended.

The practical and theoretical goals of this book will be met if readers actively participate in rapid learning system use and development. The author welcomes critical suggestions for improving and expanding this exciting new technology. Reader interest in supplying new examples and developing new rapid learning tools will be especially appreciated. Readers may wish to send suggestions and inquiries to the author *via* electronic mail (rcns_bj@mindspring.com).

Since this book covers a broad content domain, readers with limited time may wish to use book features that have been added to help them. Taking a few minutes to look over the Contents and the introductions to Parts 1 through 4 will give readers a good basis for deciding what to read and what to skip. Chapters 1 through 4 may be read in any order, although a brief reading of Chapter 1 is recommended first, because the applications in it have driven all results provided in other chapters. Some chapters in the book contain a broad range of topics and technical levels, some of which readers may wish to skip. For this reason, taking a few minutes to look over the contents and introduction to each chapter prior to reading it is also recommended.

Finally the Glossary has been included to allow reading the book without having to memorize key terms and symbols — readers who wish to save time are encouraged to use it. All terms that are underlined in this book are described in the Glossary. A preliminary scan of it is recommended to get a quick overview of book content.

ACKNOWLEDGMENTS

This book covers collaborative theoretical developments and programming efforts that have occupied many people over many years. Employees, associates, and friends of Rapid Clip Neural Systems, Inc. have kindly supplied results, graphs and plots from several studies, editorial support, and general encouragement. The author offers special thanks to Barb Ainsworth, Doug Allen, Tinley Anderson, Chris Cherry, Sally Cole, Gene Daniel, Ande DeGeer, John Edwards, John Gorman, Emory Hendrix, John Higley, Barrett Kreiner, Ray Kreiner, Mary Madden, Mike Mehrman, Bill Pearson, Don Pettit, Dick Seng, and Tyler Tatum. Members and associates of the Advanced Technology Development Center at Georgia Tech have also provided excellent advice, including Margi Berbari, Vivian Chandler, Dwight Holter, Bill Kunz, Mary Leary, and Gary Troutman.

The University of South Carolina Electrical and Computer Engineering Department supplied generous financial support throughout this project. The Process Modeling Group at the Savannah River Technology Center provided extended financial support for related preliminary research, using Department of Energy funds. Earlier research support was provided by the University of South Carolina Psychology Department and the Office of Naval Research.

The publishers of this book kindly agreed to support it from an early stage, when its future was uncertain. In addition, they have patiently allowed the book and related technology to develop at a healthy pace, and they have offered editorial advice at a good assertiveness level. Special thanks are offered to MaryAnn Cottone, Henry Flesh, Jeanne Glasser, James Harper, Robert Hauserman, Deslie Lawrence, Dianne Litwin, James Peterson, and Marjorie Spencer.

Many long days and nights have been devoted to the studies that are described in this book, by former students and colleagues at the University of South Carolina and the Savannah River Technology Center. Special thanks are due to Steve Durham, Richard Edwards, Yalin Hu, Terry Huntsberger, Sameer Joshi, Jim Laughlin, Keping Ma, Krishna Naik, Sivaram Palakodety, Donghong Qian, Chris Randall, George Weeks, and Kai Yu.

Completing this *long* project would have been impossible without strong role models, good friends, and loving relatives. I have been blessed with some marvelous role models, including R.A. Fisher, H.G. Rickover and L.L Thurstone whom I never met, along with Bill Meredith, Jerzy Neyman and Elizabeth Scott with whom I luckily studied. Friends in Berkeley, Columbia, and Atlanta — along with relatives from coast to coast — have encouraged and supported this work as well, most notably my brother Phil. Finally, this project would have been aborted long ago without my strongest role model, dearest friend, and loving wife, Carol Macera.

Although many people have devoted effort to this book, I alone have produced every mistake in it — I don't need anyone's help for that.

DISCLAIMER

All examples in this book have been performed by either early or current versions of *Rapid Learner*™ software, which have been kindly loaned by Rapid Clip Neural Systems, Inc. In addition, all the software options described in this book have either been installed or are intended for installation in *Rapid Learner*™. Hopefully, the software will be available to book readers over the Internet in the near future. However, I cannot guarantee whether or when the software will become available to book readers, nor can I guarantee the precise form that the software will take.

Atlanta, Georgia, U.S.A.
April, 1997

Concurrent Learning and Information Processing

Part 1

Rapid Learning Benefits

Part 1 describes applications that benefit from rapid learning methods and compares rapid learning with alternative methods. (Throughout this book, concurrent learning and information processing methods, along with related software and neuro-computing systems, are abbreviated as rapid learning methods and systems — all underlined terms in the book appear in the Glossary.) Chapter 1 highlights appropriate applications, ranging from simple measurement monitoring to more complicated process control and model refinement. Chapter 2 contrasts concurrent learning and information processing systems with alternative pre-programming, statistical, and neuro-computing methods. Part 1 should allow readers to recognize settings where rapid learning methods are useful alternatives — as well as complements — to conventional methods.

1

Rapid Learning Applications

	INTRODUCTION	4
1.1.	CONCURRENT LEARNING AND MONITORING	5
	1.1.1. Industrial Process Monitoring	5
	1.1.2. Structural Test Monitoring	10
	1.1.3. Health Management Monitoring	13
	1.1.4. High-Speed Monitoring	14
1.2.	CONCURRENT LEARNING AND FORECASTING	15
	1.2.1. A Commodities Forecasting Example	15
	1.2.2. High-Speed Forecasting	17
1.3.	CONCURRENT LEARNING AND CONTROL	18
	1.3.1. Missile Tracking Control	18
	1.3.2. Industrial Process Control	19
	1.3.3. Health Treatment Control	21
	1.3.4. High-Speed Control	21
1.4.	CONCURRENT LEARNING AND MODEL REFINEMENT	22
	1.4.1. A Pattern Recognition Example	22
	1.4.2. Measurement Clustering Results	23
	1.4.3. Feature Expansion Results	24
	1.4.4. Feature Removal Results	24
	1.4.5 Rapid Refinement Considerations	25
	CONCLUSION	26
	REFERENCES	27

INTRODUCTION

This chapter introduces practical application examples that are described throughout the book. The examples highlight key benefits of concurrent learning and information processing methods. (All underlined terms in this chapter and the rest of the book appear in the Glossary.) The examples are arranged in this chapter, just as they are arranged throughout this book, into three main application areas: adaptive monitoring (see section 1.1), forecasting (see section 1.2), and control (see section 1.3). This chapter describes another important aspect of concurrent learning as well: automatic model refinement (see section 1.4).

The examples have been selected to illustrate the benefits of a rapid learning approach to processing computerized data: continuously learning, without interruption, *at the same time that input-output functions are computed*. This approach differs sharply from standard alternative approaches, which require off-line estimation *prior to computing input-output functions*. Basic distinguishing benefits of concurrent learning methods include the following:

- Concurrent learning establishes input-output relationships automatically, without off-line data analysis.
- Concurrent learning adjusts to changing input-output relationships *during* input-output processing.
- Concurrent learning does not require specialized pre-programming.
- Concurrent learning keeps up with data arriving at very high rates.
- Concurrent learning utilizes and adjust thousands of input and output variables continuously.
- Concurrent learning learns relationships among hundreds of variables in less than a second.
- Concurrent learning solves a variety of large-scale, highly adaptive monitoring, forecasting, and control problems.
- Concurrent learning adapts to changing conditions much more quickly than off-line methods.
- Concurrent learning requires much less data effort than off-line analysis.

Statisticians and neuro-computing specialists often devote entire careers to solving difficult data analysis problems off-line. For them, the above claims for an automatic, real-time learning system may be hard to believe. As an initial faith offering, they might note that *animals* use concurrent learning methods to produce *precisely* the same benefits. The methods in this book simply produce these benefits with computers rather than with brains.

On the other hand, computer systems and animals that learn only in real-time cannot solve difficult data analysis problems that require careful human thought.

Such problems, which are both numerous and important, can only be solved off-line by trained specialists. These specialists will never be replaced by neuro-computing alternatives, such as the rapid learning system described in this book. The examples in this chapter represent important problems that the rapid learning system *can* solve. Yet the problems are necessarily simple and the irreplaceable human analysis element is necessarily absent from them.

1.1 CONCURRENT LEARNING AND MONITORING

This section illustrates benefits associated with rapid learning and monitoring operations with examples. One example shows how the rapid learning system identifies deviant gauge readings at once during industrial process operations (see section 1.1.1). A second example shows how the system identifies unusual fabrication results at once, without requiring off-line analysis beforehand (see section 1.1.2). A third example shows how the system may be applied in health settings (see section 1.1.3). Other examples show how the system may be useful in high-speed information processing applications (see section 1.1.4).

1.1.1 Industrial Process Monitoring

Concurrent learning and monitoring methods identify deviant instrument readings during industrial process monitoring. Measurements from many instruments, such as the temperature and tank level gauges shown in Figure 1.1.1.1, produce measurements that are immediately read by a computer. The rapid learning system then identifies deviant readings quickly, allowing operators to take corrective action at once.

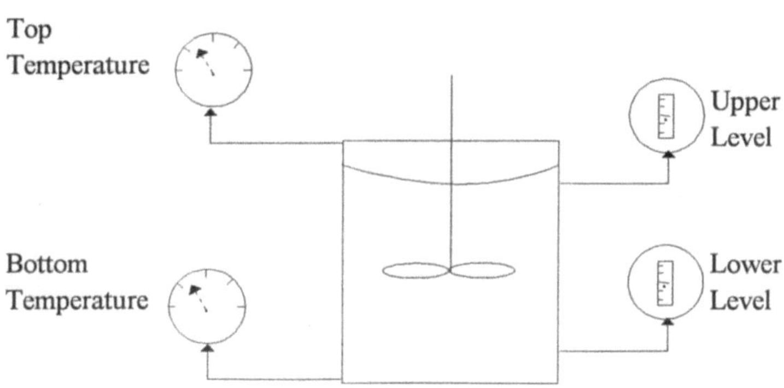

FIGURE 1.1.1.1 An Industrial Process Measurement Example
(Courtesy of Rapid Clip Neural Systems, Inc.)

Rapid learning and monitoring advantages were shown during a feasibility study at the Savannah River Technology Center [1]. Engineers at this facility tested rapid learning methods for monitoring instruments during a process that converts highly radioactive sludge into less hazardous glass logs [2-8]. When sludge and glass powder are mixed in a tank at the beginning of the process, highly sensitive instruments must continually monitor tank activity (see Figure 1.1.1.1). Occasional failure of gauges and sensors may occur in this hostile environment. Intensive manual monitoring is often employed in this and other processing settings, because conventional automated monitoring is often impractical. In particular, plant conditions vary so much that pre-programming to detect all possible modes of instrument failure under all possible plant conditions is difficult at best.

Tests at the Savannah River Technology Center showed that rapid learning methods produce simple, effective, and automatic tank measurement monitoring [2-6]. Engineers collected several data samples with faulty gauge readings, which were used to assess rapid learning monitoring capability. Measurements from one sample appear in Figure 1.1.1.2. The graph shows two sets of measurement values at each of 460 time points. One set was predicted by the rapid learning system; the other set was actually produced by the tank's upper level gauge. The rapid learning system determined each predicted value as a function of the measurements produced by the other gauges in the tank. As the graph indicates, the observed measurement values abruptly changed at time point 260.

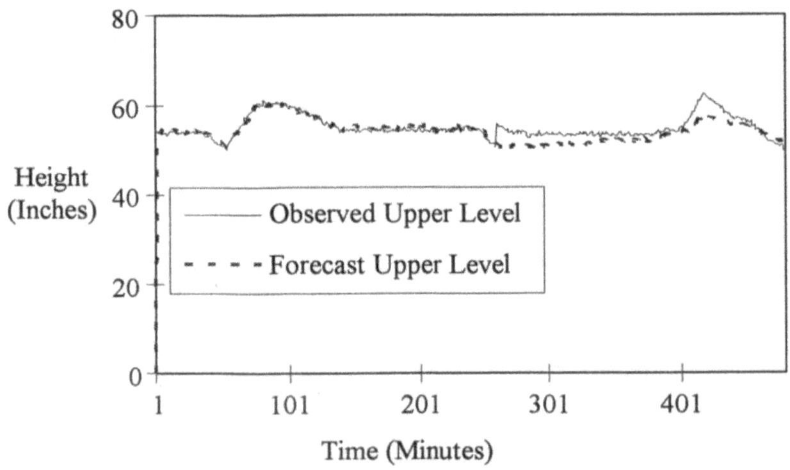

FIGURE 1.1.1.2 Observed And Predicted Level Gauge Measurement Values (Courtesy of Rapid Clip Neural Systems, Inc.)

When the forecast values did not change, the rapid learning system immediately and accurately interpreted the resulting discrepancy as a gauge failure.

Besides showing effective monitoring capability, this experiment demonstrated fast learning capacity. As shown at the beginning of Figure 1.1.1.2, the rapid learning system began to forecast the measurements of the upper level gauge accurately, within the first few time points. The system also learned automatically, without requiring any user-supplied measurement functions. Furthermore, the system was able to update learning quickly, whenever relationships among measurements changed.

For this kind of monitoring application, the rapid learning system provides the following benefits:

- It automatically learns to predict each measurement during processing.
- It predicts each measurement as a function of all other input measurements at the same point in time.
- It identifies unusual values by comparing each predicted measurement with each observed measurement.

Rapid learning and monitoring methods offer benefits beyond those required for this study. In this study, at one minute intervals the rapid learning system predicted measurement values for each gauge based on a few other concurrent gauge measurements. But the system can also function in the same way if many more measurements arrive far more frequently (see section 2.4). In addition, it can use historical measurements as well as concurrent measurements to forecast measurements well into the future (see section 7.2). The system is also designed to identify a variety of linear and nonlinear features in order to meet a variety of application needs (see Chapter 6).

Off-line data analysis requires time to collect and analyze training data before useful monitoring can begin. Off-line methods are also designed to identify prediction functions for each measurement separately. Identifying these functions takes valuable analysis time. Furthermore, this time-consuming identification process must be repeated whenever process conditions change. In contrast, the rapid learning system uses many simple relationships to predict all measurements at once, and it learns the relationships automatically.

As a second example associated with the above hazardous waste process, stainless steel canisters are filled with molten glass mixed with radioactive waste as part of the process. After the canisters are filled, caps are welded as shown in Figure 1.1.1.3. During the two seconds that cap welds are formed, 12 force, motion, and electricity measurements are made every 25 milliseconds. The rapid learning system assesses cap weld quality from these measurements, it identifies

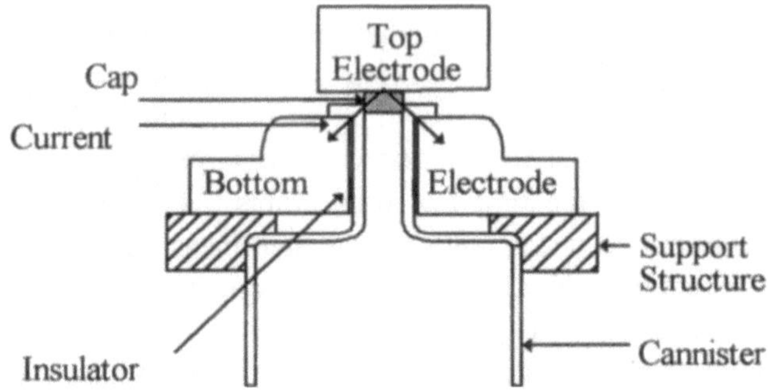

FIGURE 1.1.1.3 A Quality Monitoring Application
(Courtesy of Rapid Clip Neural Systems, Inc.)

unusual measurement profiles for each weld, and it identifies unusual measurement trends from weld to weld. Once unusual discrepancies are identified, inspectors are alerted. Inspectors are also alerted if weld profile trends are unusual, indicating that welding equipment is beginning to break down.

To monitor weld quality, the rapid learning system utilizes weld quality functions that have been identified from previous weld studies. Such functions reduce several thousand measurements to several key measurement functions, known as <u>features</u>. The system then performs the following operations immediately after each weld:

- It computes all feature values.
- It predicts each current feature value from other current feature values.
- It compares each predicted feature value to previously learned feature variation statistics to produce deviance statistics.
- It sounds alarms if the deviance statistics are unusually large.
- It produces an overall deviance statistic for the current feature profile and produces an alarm if it is unusually large.
- It updates all learned statistics that are necessary for monitoring prior to receiving data for the next weld.

Figure 1.1.1.4 shows weld amperage profiles that have been used to predict canister weld quality. Frame (a) shows a typical moderate duration profile (over 1.5 seconds) among 22 that were obtained, frame (b) shows one of 8 short-duration profiles, frame (c) shows one of 8 long-duration profiles, and frame (d) shows an unusual profile. Each of these 39 welds was burst tested and then analyzed off-line using statistical methods, to identify feature functions that accurately predict weld strength [7]. Off-line statistical methods also identified feature value regions within which weld strength was acceptably high.

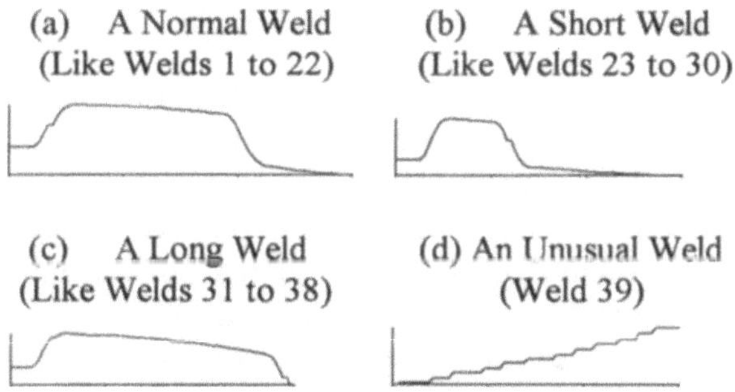

FIGURE 1.1.1.4 Concurrent Learning and Monitoring Assessment Data (Courtesy of Rapid Clip Neural Systems, Inc.)

To illustrate concurrent learning and profile monitoring for this application, the rapid learning system processed the 39 measurement profiles like those in Figure 1.1.1.4 sequentially, beginning with 22 like the (a) weld, followed by 8 like the (b) weld, followed by 8 like the (c) weld and ending with the (d) weld. During each processing trial the deviance of the concurrent profile from profiles that had been learned up to that point was assessed, and learning was updated to include the concurrent profile. No prior feature learning was utilized in this experiment. Instead, the raw measurements were read and relationships among them were learned starting from trial one, with no user-supplied initial information.

As expected, profile deviance values were alarmingly high immediately after trials 23, 31 and 39. Such values were significantly high relative to other deviance values, because the profiles at each of these time points differed greatly from those that had been learned up to that time point.

Rapid learning system capacity to identify such abnormal profiles does not distinguish it from alternative methods. Off-line systems may be able to perform just as accurately when relationships among measurements are stationary (see section 2.2). Furthermore, off-line systems may be adequate even when relationships are non-stationary, provided that measurements arrive slowly. For example, rapid adaptation speed is convenient but not essential for monitoring the weld profiles in Figure 1.1.1.4, because several hours are available between welds to update learning off-line. The distinguishing benefit is the rapid learning system capacity to update learning very quickly, in applications when information is arriving very quickly (see section 2.4). For example, rapid learning software can adaptively monitor profiles made up of several hundred measurement features (see section 2.4.2), even when profiles arrive every second. Furthermore, special-purpose rapid learning hardware can monitor measurements arriving much more quickly (see 2.4.1).

1.1.2 Structural Test Monitoring

Concurrent learning and monitoring systems offer major improvements in the complex testing of aircraft components. Aircraft structural testing often results in damage to expensive design prototypes. Damage results because all available strain gauge measurements cannot be monitored precisely during testing.

In a case study conducted for a large aircraft manufacturing company, the rapid learning system demonstrated the advantages of learning during structural testing [9]. Without relying on prior data analysis, the system identified an impending failure region in a heavily stressed tail section. The region was identified early enough so that damage could have been avoided if the system had been used.

The tail section used for the case study is illustrated in Figure 1.1.2.1. Hydraulic actuators attached to the tail section, located at the circles in the figure, applied loads to simulate in-flight maneuvers during several test trials. Loads were raised from low levels to high levels during each trial. Strain gauge readings at several time points during each trail were provided. Measurements were supplied in the form of one record per time point, with each record containing all gauge readings at that time point.

The rapid learning system performed the following operations in real time:

- It learned the relationships among incoming gauge measurements during the first few measurement time points.
- It predicted measurement values for each gauge at each time point.

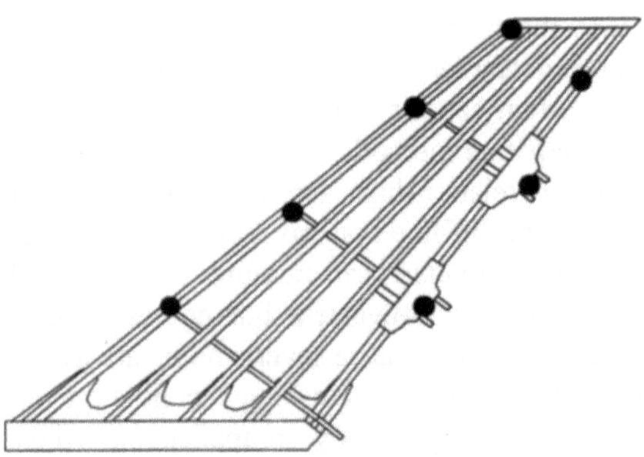

FIGURE 1.1.2.1 An Aerospace Structural Test Example
(Courtesy of Rapid Clip Neural Systems, Inc.)

- It monitored measurement records for discrepancies between observed and predicted values at each time point.
- It updated learned relationships continuously to increase monitoring precision.
- It identified the region of developing damage prior to test article failure.

The rapid learning system received one set of 197 strain gauge measurements at each time point during three trials (Figure 1.1.2.2). Learning without relying on previously gathered training data, the system accurately predicted each gauge value as a function of the other 196 values after the first four time points shown in Figure 1.1.2.2. In order to monitor at each time point, a deviance value was computed between each observed and predicted gauge value. If deviance values exceeded a pre-determined criterion value, corresponding gauges were classified as deviant and their locations appeared in a deviant gauge location plot.

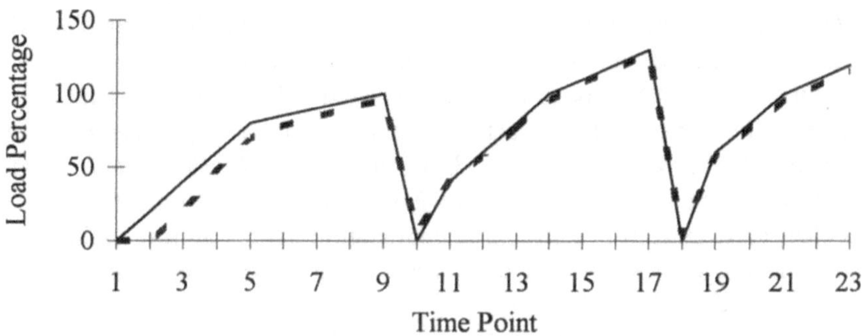

FIGURE 1.1.2.2 Load Versus Time During a Structural Test
(Courtesy of Rapid Clip Neural Systems, Inc.)

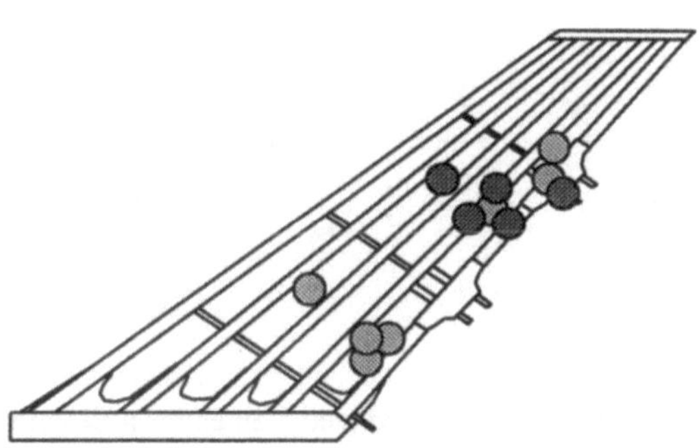

FIGURE 1.1.2.3 A Strain Gauge Monitoring Display
(Courtesy of Rapid Clip Neural Systems, Inc.)

Figure 1.1.2.3 shows selected deviant gauge location plots that were prepared by the rapid learning system for the case study. Each plot displays deviant gauge locations according to the following color scheme:

- Each dark spot corresponds to a gauge that has just become deviant.
- Each light spot corresponds to a gauge that first became deviant prior to the current time point.

In an operational setting, test engineers can see displays like Figure 1.1.2.3 every few seconds. Test engineers can use such displays to scan a large number of data channels rapidly and identify regions of unusual activity. Looking at these displays allows them to decide which gauges should be examined further and when to terminate a test.

As part of the case study, test engineers were shown displays like Figure 1.1.2.3 from the middle of trial 1 onward. When they saw Figure 1.1.2.3 after viewing a series of uneventful displays, they said they would have examined deviant gauges in the figure before further continuing the structural test.

Figure 1.1.2.4 shows observed and predicted values for one of the deviant gauges that the test engineers selected. The Figure 1.1.2.4 plot shows a developing departure between observed (solid) and predicted (dashed) gauge strain at the time point when Figure 1.1.2.3 was generated (point 19 on the plot). This was a clear early indicator of structural degradation, in that all test engineers participating in the case study said they would have discontinued the test at that point.

As it happened, test engineers did not have the rapid learning system at their disposal during the test. Since they had no indication that structural degradation was occurring, the test was continued well beyond time point 3. After several

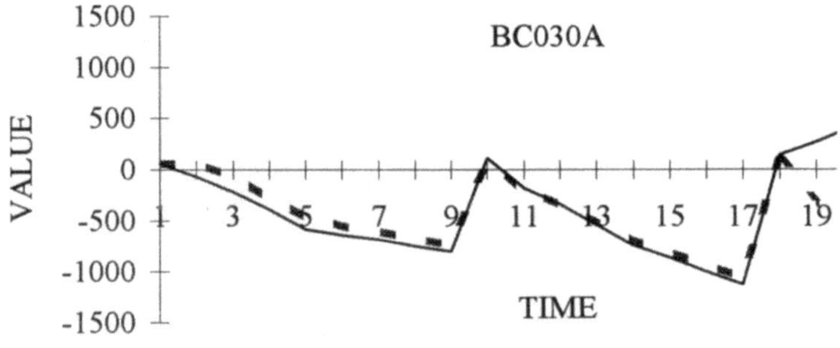

FIGURE 1.1.2.4 Observed and Predicted Values for a Selected Strain Gauge (Courtesy of Rapid Clip Neural Systems, Inc.)

more time points, the tail section visibly broke in the precise region that would have been identified by the system at time point 19.

This case study showed that the rapid learning system adds monitoring value. The system has the capacity to predict and monitor instrument readings without prior modeling, because it learns complex relationships *during* data entry. As a result, the system provides a useful solution for structural test monitoring problems.

1.1.3 Health Management Monitoring

Health management monitoring systems such as the one shown in Figure 1.1.3 are becoming increasingly common. These systems range from devices for monitoring vital signs (temperature, pulse rate, etc.) to hand-held devices used by home care nurses for recording health behavior. Typically, monitoring deviance alarms are sounded and messages are generated if values of recorded measurements exceed pre-set tolerance band limits.

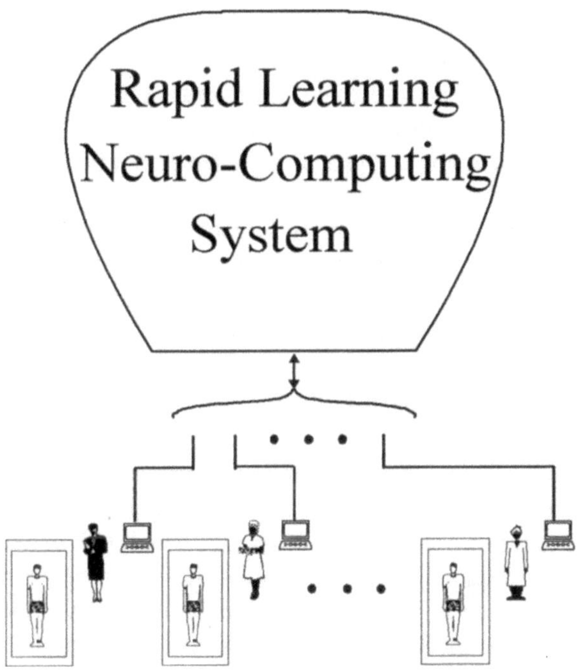

FIGURE 1.1.3 A Health Management Application
(Courtesy of Rapid Clip Neural Systems, Inc.)

The key problem with typical health management monitoring systems is that tolerance bands tend to be too wide for effective monitoring. The reason is that pre-programmed bands must be wide enough to cover a broad range of values that occur among healthy people. Furthermore, vital sign and health behavior measurements tend to depend heavily on each other and on each patient's health history. For example, a hip replacement patient typically receives pain medication during her stay, depending on her own progress. No pain medication on the third day after surgery might be normal for hip replacement patients in general. However, substantial pain medication might be normal for patients who have had complications, which might be indicated by measurements such as daily blood transfusions. Pre-programmed tolerance bands must be very wide to account for such a wide variety of complications.

The rapid learning system can precisely monitor health measurements, because it can quickly learn baseline tolerance bands for individual patients [10]. These tolerance bands are more precise than pre-set tolerance bands, because the system creates each band as a conditional function of other measurements. The system can also track patients over time effectively, taking into account individual case histories. For example, the system has the capacity to perform the following steps in real time, during each patient's hip replacement visit:

- It can identify unusual entries in the patient's admission record.
- It can identify unusual treatment levels during each day of admission as a function of earlier entries for the patient.
- It can update learning at the end of the patient's admission.

Precise monitoring is possible because each measurement is evaluated relative to previous measurements for the patient. Also, precision increases as more patients are monitored since the system updates learned correlations among monitored measurements as a matter of course.

1.1.4 High-Speed Monitoring

Rapid learning and monitoring benefits tend to increase with increasing data arrival rates, because changing relationships may develop more quickly, off-line alternatives may adapt more slowly, and costs associated with undetected problems may mount up more quickly (see section 2.4.5). Increased benefits especially result when added value is proportional to monitoring and response frequency, for example in real-time market trading applications (see section 1.2). If relationships among prices are changing from time to time then direct payoff benefits can be achieved by keeping up with such changing relationships. Furthermore, if prices are monitored every minute, resulting benefits can be far higher than if prices are monitored only every day (see section 2.4.5), simply because many more trading opportunities will emerge.

Because rapid learning and monitoring benefits are tied to data arrival rates in this way, the highest benefit potential lies in applications where information arrives at very high-speeds. Monitoring high-speed, rotating electrical equipment is thus a good application, and monitoring telecommunications activity for unusual behavior is a still better application. As a telecommunication monitoring example, so-called search engines on the Internet may benefit from rapid learning methods that identify new information access patterns as they develop. Once such new patterns have been identified, new data access channels and schemes can be opened and adjusted to handle new data traffic patterns. Likewise, memory access schemes can be adjusted as changing input-output patterns are identified during computer use.

Accordingly, rapid learning system development has focused on achieving concurrent learning and monitoring in very fast data arrival applications. While rapid learning software can keep up with high-speed data arrival, rapid learning hardware can keep up with very high-speed arrival. For example, rapid learning software can adaptively monitor each of several hundred gauge measurements arriving at the rate of one set per second, but rapid learning digital hardware is being developed to operate 20,000 times faster than rapid learning software. Furthermore, rapid learning analog hardware is being developed to operate 1,000 times faster than rapid learning digital hardware (Section 2.4.1).

1.2 CONCURRENT LEARNING AND FORECASTING

This section illustrates benefits associated with rapid learning and forecasting operations, beginning with a commodities forecasting example (see section 1.2.1). The example contains most of the basic elements that are found in many rapid forecasting applications. The section also includes discussions of other applications where very fast concurrent learning and forecasting are required (see section 1.2.2).

1.2.1 A Commodities Forecasting Example

Figure 1.2.1 shows Standard & Poors commodities index values on a minute-by-minute rapid learning system display [11]. The thin black plot shows actual values up to 11:07 A.M., the current time of day. The thick black plot beginning at 11:07 includes forecast values 5 minutes into the future. The upper and lower dashed plots beginning at 11:07 are upper and lower tolerance bands for the forecast values, beginning at 11:07. The thick and dashed lines shown before 11:07 are forecast and tolerance bands that were previously available at 10:49 and 10:59.

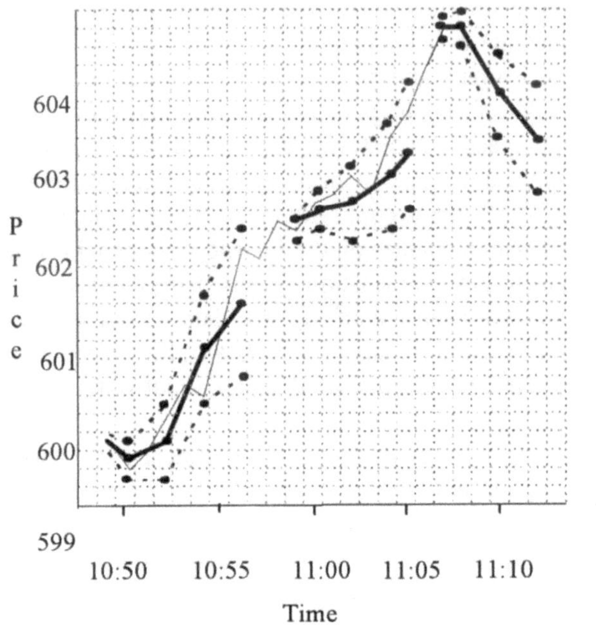

FIGURE 1.2.1 A Commodities Forecasting Setting
(Courtesy of Rapid Clip Neural Systems, Inc.)

At 1-minute intervals, the rapid learning system performs the following forecasting operations for data arriving in this way:

- It receives new Index values for the current minute, along with a variety of other measured values for the current minute.
- It computes and plots Index forecast and tolerance band values for the current minute.
- It computes and plots forecast and tolerance band values for other measured values at the current minute.
- It updates each learned forecasting function.
- It updates each learned tolerance band function.

The results shown in Figure 1.2.1 are valuable to commodities traders who buy and sell the Index on a minute-by-minute basis. Typically, traders make very short-term decisions such as buying shares and selling them 15 minutes later or selling shares and buying them 15 minutes later. As an aid in making these decisions, each Figure 1.2.1 forecast telescope (forecast and corresponding tolerance band) provides useful information to the trader. For example, the historical telescope at 10:49 indicates that a buying transaction might have been profitable at that time; the historical telescope for 11:00 indicates that holding the Index shares bought at 10:49 instead of selling them back would have been profitable at that time; and the current telescope at 11:07 indicates that a current selling transaction might be profitable.

A preliminary study has shown that the rapid learning system can provide useful information to commodities traders over and beyond typical information that they use [11]. Unlike off-line alternatives, on a minute-by-minute basis the system can quickly forecast each of many measured values as a function of current and recent values for all other measured values. Furthermore, the system can update learned relationships among all such measured values continuously, so that each forecast is based on up-to-the-minute learning. This capacity to learn and forecast concurrently is especially important in forecasting applications such as commodities trading, where relationships among measurements may change very quickly and dramatically.

1.2.2 High-Speed Forecasting

The rapid learning system capacity to forecast adaptively at higher than minute-by-minute rates is useful in a variety of trading and other applications. Increasingly, large trading firms are making automatic computerized trading decisions based on arbitrage (deviance values between observed and predicted prices), within seconds rather than minutes. As in the Standard & Poors setting, relationships among measurements change rapidly in arbitrage settings as well. As a result, the system offers higher precision trading than off-line alternatives, which are based on static tolerance bands that can only be updated occasionally.

The value of rapid learning relative the value of off-line alternatives increases with measurement arrival frequency (see Section 2.4.5). For example, a reasonable method for forecasting a commodities index 30 time points into the future might involve fitting curves to the index values for the most recent 90 time points and then projecting them 30 time points into the future. If time points are one day apart, standard statistical software can easily be used to obtain the desired forecast every evening, between trading sessions. Off-line statistical analysis can also provide updated forecast learning, based on discarding training measurements for 90 days ago and replacing them with measurements for the most recent day. If time points are an hour apart, however, updating between each time point would be more difficult. If time points are a minute apart and monitoring each of 500 index component prices is necessary, the statistical approach will be impractical. By contrast, the rapid learning system can easily keep up with such forecasting requirements, even if time points are a few seconds apart (Section 2.4.1).

Rapid learning software and hardware can forecast and update learning more rapidly than every few seconds, for a variety of more time-demanding applications. Forecasting associated with high-speed mechanical equipment like electric power generators is one promising application. Being able to forecast important variables such as acute power demand is also an essential element of rapid control (see next section). Furthermore, being able to adapt high-speed forecasts to rapidly changing relationships increases forecasting precision, while avoiding costly down-time associated with off-line data analysis (see Chapter 2). As in high-speed monitoring, the best potential value for rapid learning forecasting is in electronic applications.

1.3 CONCURRENT LEARNING AND CONTROL

This section illustrates benefits associated with rapid learning and control, beginning with a missile tracking control example (see section 1.3.1). The example contains most of the basic elements that are found in many rapid control applications. The section also describes a more challenging closed-loop industrial process control example (see section 1.2.2). In addition, the section introduces related health applications (see section 1.2.3), as well as applications requiring very rapid learning (see section 1.2.4).

1.3.1 Missile Tracking Control

Figure 1.3.1 shows observed and forecast position coordinates for a rapidly moving missile from a simulation experiment [12]. At each time point, the rapid learning system performed the following steps during the experiment:

- It received new position information for the missile.

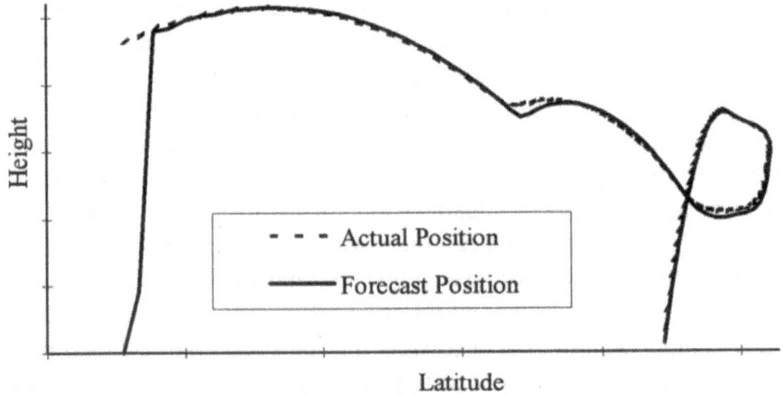

FIGURE 1.3.1 A Missile Tracking Control Setting
(Courtesy of Rapid Clip Neural Systems, Inc.)

- It forecast the next missile coordinates as a function of current and recent position information.
- It provided output forecast values and tolerance bands for missile tracking control.
- It updated learned relationships among position information measurements.

As Figure 1.3.1 shows, the rapid learning system was able to track and forecast missile position after only a few time frames. Furthermore, it was able to keep track of missile position during several abrupt excursions. Alternative systems can keep track of missile position just as quickly as the rapid learning system, provided that all missile excursions can be anticipated, analyzed and preprogrammed. For example, when missile movement is purely ballistic without deviant excursions, pre-computed extrapolation functions can track missile position very effectively.

The rapid learning system takes a different tracking approach than alternative methods take. Instead of attempting to pre-program all excursions, it continuously learns to use recently measured position measurements for forecasting next position coordinates. Since it can use many such recent measurements to forecast positions very quickly, the system can forecast next position coordinates very precisely. Moreover, it can continue to forecast precisely even during totally unexpected excursions, as Figure 1.3.1 illustrates.

1.3.2 Industrial Process Control

Practical control problems range from relatively simple to quite complex. One complicating factor is the potential for the controlling process itself to influence future measurements (see section 8.1.2). For example, the above missile control problem is relatively simple to study and solve, provided that missile position excursions do not depend upon missile tracking control. Otherwise, the problem requires more complicated models and methods. In particular, suppose that the missile takes evasive action in reaction to tracking accuracy measurements that it monitors. In that case, relevant rapid learning studies will require either tracking an actual missile taking evasive action or simulating such evasive action from trial to trial.

Figure 1.3.2 illustrates a typical adaptive control setting where the controlling process influences future measurements. The figure describes liquid being sent to a mixing tank through an input line containing a controlled input valve. The control system for such a valve may receive a variety of measured inputs, such as the level and temperature measurements that are shown in the figure. Once the control system receives such measurements, it sends signals to open or

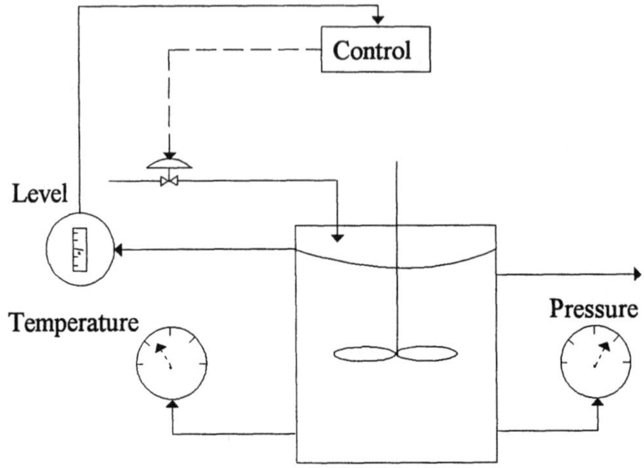

FIGURE 1.3.2 An Adaptive Control Setting
(Courtesy of Rapid Clip Neural Systems, Inc.)

close the valve accordingly. For example, the system might send a closing signal to the valve if measured tank level exceeds a pre-set value, and it might send an opening signal to the valve if measured tank level is less than another pre-set value. The control system directly affects future level measurements, in that valve closing signals will tend to lower future level readings and valve opening systems will tend to raise future readings. The control system may affect other measurements besides level as well. For example, closing the control valve may tend to produce warmer input liquid because the liquid may be passing over upstream heaters more slowly.

The rapid learning system can solve concurrent learning and control problems like the one shown in Figure 1.3.2 (see section 8.1.2). The system performs the following steps at every time point:

- It receives current input measurements from the control system.
- It provides a valve control signal that depends on tank level.
- It predicts tank level.
- It updates weights for predicting tank level from other measurements.

Rapid learning and control systems can maintain tank level well during system excursions, even when relationships among measurements are changing during the process.

In contrast with rapid learning and control systems, conventional methods typically adjust control functions off-line. Off-line adjustments are required prior to initial installation and whenever measurements among measurements change. Such off-line calibrations are required among standard PID controllers (see section 2.1.1, 8.1.2), as well as adaptive fuzzy and neural controllers [13, 14]. Off-line calibration shares the same liabilities for control as off-line training for monitoring and forecasting: off-line analysis takes time, reducing adaptation speed, and requires manual effort, raising overall expense. Off-line control calibration has one added liability that can be costly in industrial process settings: the process must often be shut down during calibration in order for necessary analyses to be performed. As a result, rapid learning and control offers real savings over conventional methods along several lines.

1.3.3 Health Treatment Control

The Figure 1.1.3 health management setting that was introduced earlier also describes a rapid learning and health control setting. Expert advisory systems are available for a variety of applications including pain treatment [10]. In their standard form, health advisory systems provide treatment suggestions based on patient information and pre-programmed rules. For a pain treatment advisory system, a health professional may enter a patient's background information, vital signs, and information about the patient's pain. Once the system has received the information, it supplies a list of suggested treatments that have been pre-programmed to fit the patient's profile. Each option in the list may include a treatment effectiveness estimate.

An alternative rapid learning and control system can operate initially just as a standard advisory system, but it can use treatment effectiveness ratings to improve advice as it gains experience [10]. After each treatment, the patient or treatment provider enters an effectiveness rating into the system. The system then learns from the event by adjusting connection weights. The result of a low effectiveness rating is to reduce future effectiveness estimates of that treatment among similar patients. The result of a high effectiveness rating is the opposite.

The overall effect of concurrent learning advisory systems is thus to control the delivery of advice, hence actions that are taken based on that advice, to produce gradual and automatic improvement based on experience.

1.3.4 High-Speed Control

Because of its rapid learning capacity, the rapid learning system is able to control adaptively under rapidly changing conditions. For example, standard rapid learning software can easily keep up with missile tracking information in the form of 64 measurements arriving every 65 milliseconds, and far faster rapid learning hardware may be used if faster adaptive control is required (see Section 1.4.1). As in high-speed monitoring and forecasting applications (see Sections 1.2 and

1.3), the value of rapid learning software relative to off-line alternatives increases in such high-speed settings, because adaptation speed differences between rapid learning and alternative approaches become more pronounced.

Just as in high-speed monitoring and forecasting applications, the highest potential value for rapid learning control is in electronic applications. Rapid learning advantages for Internet and computing monitoring and forecasting applications that were outlined in Sections 1.1 and 1.2 apply for control as well. Rapid learning advantages are even more pronounced for electronic control, because taking operational electronic systems off line for control calibration can be very expensive. Indeed, among all rapid learning applications, adaptive electronic control have the highest potential value.

High control speed is also important in batch processing applications, even when fast response time may not be necessary. For example, health treatment advice control (see Section 1.3.3) turnaround of a few seconds per case is sufficiently fast. However, if a single computer must receive inquiries from a worldwide network, being able to process each inquiry in real time will require very high processing speed.

1.4 CONCURRENT LEARNING AND MODEL REFINEMENT

This section reviews a case study that illustrates how and why rapid learning and model refinement operations are performed (see section 1.4.1). The case study contains many of the basic elements that are found in typical model refinement operations: combining redundant measurements (see section 1.4.2), introducing new measurements (see section 1.4.3), and removing unnecessary measurements (see section 1.4.4). General refinement concerns that led to the rapid learning system's modular design are also described (see section 1.4.5).

1.4.1 A Pattern Recognition Example

Figure 1.4.1 shows 16 patterns that have been analyzed by the rapid learning system in a study to simulate recognition of "noisy" (randomly distorted) visual data [15]. In the study, noisy counterparts to each of the 16 patterns were generated in such a way that any of the 49 pixels in each pattern were randomly switched from light to dark or dark to light. In many related pattern recognition applications such as textile quality assessment, a large number of individual measurements are available as inputs — often too many to be interconnected in a rapid learning system. For example, the number of possible pair-wise connections among pixels in a 1,024 by 1,024 visual array is 524,800. Although the rapid learning system can easily operate with pixel arrays of this size, pixel measurements must first be combined into a much smaller number of features before rapid learning can occur.

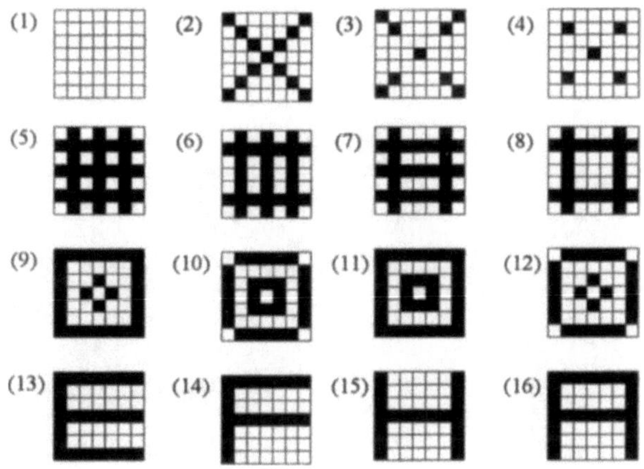

FIGURE 1.4.1 A Pattern Recognition Example
(Courtesy of Rapid Clip Neural Systems, Inc.)

Visual patterns, sound patterns, and other data arrays often contain a great deal of redundant information. Often, redundant information can be identified and dealt with in a way that can greatly reduce network size and increase network operating speed. For example, within each of the 16 patterns in Figure 1.4.1 the upper left pixel has the same shade as the upper right pixel. Thus, all of the information in the 16 upper left pixels that distinguishes the patterns is precisely the same as the information in the 16 upper right pixels. If redundancies such as these can be identified, then redundant information can be reduced to produce a much smaller number of sufficient features.

Neuro-computing systems can operate as accurately with sufficient features as with the original measurements, but much more quickly and with far lower memory requirements. For example, only 12 sufficient features are needed to distinguish the 16 Figure 1.4.1 patterns, instead of the 49 individual pixel measurements. The resulting reduction in interconnections for pattern recognition in this case, from 2,080 to 378, produces proportionate computer storage and response time savings (see section 2.4.1, 2.4.2).

1.4.2 Measurement Clustering Results

Results from this case study showed that the rapid learning system also identifies and clusters redundant information accurately. For example, Figure 1.4.2 shows the clusters of redundant pixels that the system identified from noisy data based on the Figure 1.4.1 patterns. These are precisely the same clusters that were identified from linear algebra analysis of the noise-free Figure 1.4.1 patterns themselves.

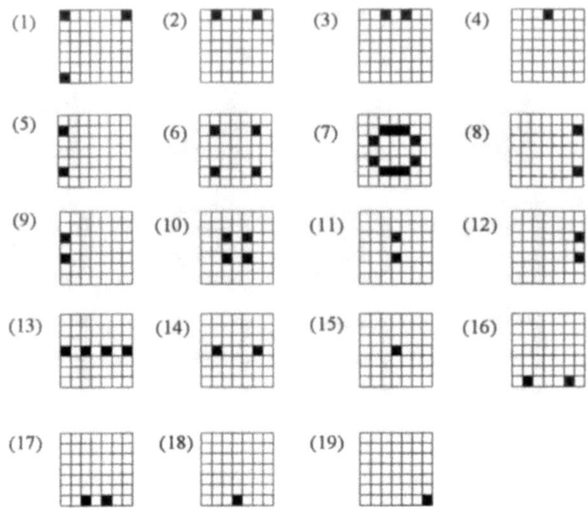

FIGURE 1.4.2 Rapid Clustering Results
(Courtesy of Rapid Clip Neural Systems, Inc.)

1.4.3 Feature Expansion Results

Whenever computer storage space or concurrent operation time permits, the rapid learning system has the capacity to use new measurement features in order to improve prediction accuracy. Using new features can reduce tolerance bandwidths and increase correct classification rates, whenever added predicted measurements depend on the new features. Expanded features may take the form of historical features, product features and other measurement functions.

For the visual pattern recognition case shown in Figure 1.4.1, new features made up of cross-products among the 12 features shown in Figure 1.4.2 were used in addition to the 12 features themselves. The rapid learning system computed all 78 such cross-products and then verified that pattern recognition performance increased after the cross-products were included.

1.4.4 Feature Removal Results

The rapid learning system has the capacity to remove unnecessary measurement features that have no effect on prediction accuracy. The rapid learning procedure for removing unnecessary features is based on well-known step-wise predictor removal regression procedures (see section 8.2). Feature removal is useful be-

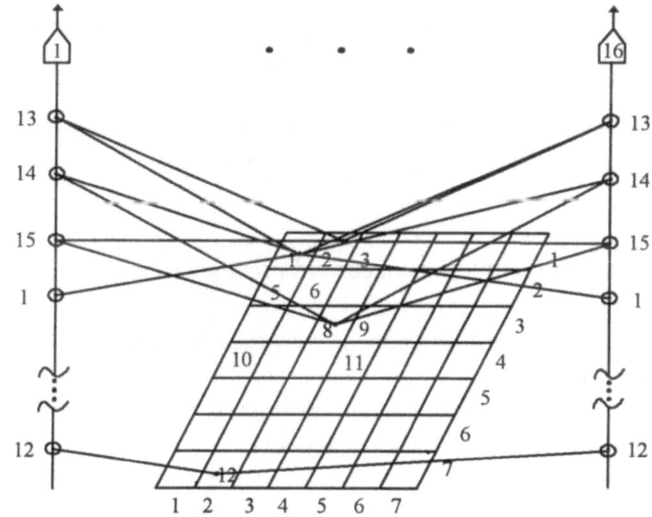

FIGURE 1.4.4 Feature Removal Results
(Courtesy of Rapid Clip Neural Systems, Inc.)

cause fewer features require less storage, permit faster concurrent operation and leave room for feature set expansion.

First-order and second-order feature functions from the Figure 1.4.1 case study were used to demonstrate rapid learning feature removal effectiveness. Rapid learning feature removal methods were applied to connection weights based on the 12 combined first-order features (see section 1.4.2), along with their 78 cross-products (see section 1.4.3). After unnecessary feature removal was complete, the 15 features shown in Figure 1.4.4 remained. The removal process was optimal in this case, in that 15 features such as those in Figure 1.4.4 are necessary and sufficient for solving the Figure 1.4.1 pattern recognition problem.

1.4.5 Rapid Refinement Considerations

The rapid learning refinement sub-system has been designed in a way that reinforces the distinguishing concurrent learning and information processing feature: its capacity to learn *during* information processing, as quickly as information arrives. The rapid system separates learning into two distinct operations that can be performed in parallel: connection weight updating and model refinement. Connection weight updating, which is very fast, is accomplished between the arrival of each measurement record (see Chapter 5). Rapid learning refinement operations have been carefully formulated to be fast (see section 8.2), but they take more time than concurrent weight updating on conventional computers. For this reason, rapid learning refinement operations have been designed to proceed separately from concurrent weight updating. In this way, connection weights can be concurrently updated between each measurement record, while refinement operations proceed more deliberately.

For the above refinement example, the rapid learning system can perform pattern recognition and connection weight updating concurrently, while performing model refinement at the same time:

- For each data record, the concurrent process can identify the pattern type if it is missing.
- For each data record, the concurrent process can update learned connection weight values.
- At any point in time, the concurrent process can send learned connection weights to a separate refinement process.
- The refinement process can then cluster redundant variables, add new features and remove unnecessary features while the concurrent process is operating.
- The refinement process can then reconfigure the concurrent process accordingly, for more efficient and/or more precise operation.
- Concurrent prediction and learning operations, as well as occasional refinement operations, may proceed in this way continuously, if necessary.

Concurrent connection weight updating and occasional model refinement can proceed continuously, but without interruption.

Along with this general scheme for concurrent operation and rapid refinement, the rapid learning system uses a variety of efficient, parallel processing schemes to accomplish refinement quickly. These schemes rely solely on connection weights to identify redundant and unnecessary measurements, thus assuring concurrent process and refinement process separability (see section 8.2).

CONCLUSION

Future Directions

This chapter highlights some key case studies that were performed to demonstrate current rapid learning monitoring, forecasting and control capacity. Other case studies will be added as rapid learning capacity expands and new applications emerge. Planned efforts will produce very high-speed learning for new computing and telecommunications applications, along with enhanced refinement capacity for new image processing applications. These planned improvements and case studies will be described later in this book and in future publications.

Summary

This chapter uses a range of examples to demonstrate key benefits of rapid learning methods. The key feature of the rapid learning system is its capacity to learn *during* information processing. The main resulting benefit is its capacity to perform monitoring, forecasting, and control with optimal precision and without interruption, even when new relationships among measurements are developing quickly.

The examples in this chapter are a small sample from a broad application field, including these:

- Industrial process, structural test, and health management monitoring
- Commodities forecasting
- Missile tracking, industrial process, and health treatment control

Since key rapid learning system benefits result from its fast adaptability, its best potential lies in non-stationary and high-speed monitoring, forecasting, and control applications. These include computer network control, telecommunication monitoring, and other applications where measurements are sent directly to computers in real-time.

REFERENCES

1. R.J. Jannarone, "Concurrent Information Processing, I: An Applications Overview," *Applied Computing Review,* Vol. I, pp. 1-6, 1993.
2. G.E. Weeks, W.E. Daniel, R.E. Edwards, Jr., S.N. Palakodety, S.S. Joshi, D. Qian, & R.J. Jannarone, "Holledge Gauge Failure Testing using Concurrent Information Processing Algorithm," *Proceedings of the International Topic Meeting on Nuclear and Hazardous Waste Management*, American Nuclear Society, Inc., La Grange Park, IL, Vol. 3, pp. 2261-2268, 1996.
3. "Industrial Instrument Monitoring," *RCNS Technical Report Series*, No. APP95-06, Rapid Clip Neural Systems, Inc., Atlanta, GA, 1996.
4. S. N. Joshi, *A General-Purpose Concurrent Information Processing Prototype*, Unpublished Masters Thesis, University of South Carolina, 1994.
5. S. S. Palakodety, *A Special-Purpose Concurrent Information Processing System*, Unpublished Masters Thesis, University of South Carolina, 1994.
6. Y. Hu, "Automated Real-Time Neural Computing for Defense Waste Processing," *Proceedings of the International Topic Meeting on Nuclear and Hazardous Waste Management*, American Nuclear Society, Inc., La Grange Park, IL, Vol. 1, pp. 534-540, 1992.

7. S.D. Durham, Z.Q. Li, R.J. Jannarone, R.E. Edwards, Jr., & J.S. Lyons, "Welder Parametric Study: Preliminary Statistical Analysis," SCUREF — Westinghouse Savannah River Laboratory Technical Report, April 1992.
8. T.L. Huntsberger & N. Navlani, "Welder Parametric Study: Nonparametric Neurocomputing Analysis," SCUREF — Westinghouse Savannah River Laboratory Technical Report, April 1992.
9. "Structural Testing of Aircraft Components," *RCNS Technical Report Series*, No. APP96-01, Rapid Clip Neural Systems, Inc., Atlanta, GA, 1996.
10. "Health Monitoring, Management, and Advisement," *RCNS Technical Report Series*, No. APP96-03, Rapid Clip Neural Systems, Inc., Atlanta, GA, 1996.
11. "A Commodity Price Forecasting Example," *RCNS Technical Report Series*, No. APP96-02, Rapid Clip Neural Systems, Inc., Atlanta, GA, 1996.
12. "Vehicle Tracking Applications," *RCNS Technical Report Series*, No. APP96-04, Rapid Clip Neural Systems, Inc., Atlanta, GA, 1996.
13. K.J. Aström & T.J. McAvoy, "Intelligent Control: an Overview and Evaluation," in D.A. White & D.A. Sofge (Eds.), *Handbook of Intelligent Control*, Van Nostrand Reinhold, New York, 1992.
14. B. Kosko, *Neural Networks and Fuzzy Systems*, Prentice-Hall, Englewood Cliffs, NJ, 1992.
15. Y. Hu, *Concurrent Information Processing with pattern Recognition Applications*, Unpublished Doctoral Dissertation, University of South Carolina, 1994.

2

Contrasts with Alternative Methods

	INTRODUCTION	30
2.1.	RAPID LEARNING VERSUS PRE-PROGRAMMING	30
	2.1.1. Pre-Programmed Prediction	31
	2.1.2. Rapid Learning as an Alternative	32
	2.1.3. Rapid Learning as a Complement	36
2.2.	RAPID LEARNING VERSUS STATISTICAL ANALYSIS	37
	2.2.1. Statistical Analysis and Prediction	38
	2.2.2. Rapid Learning as an Alternative	40
	2.2.3. Rapid Learning as a Complement	42
2.3.	RAPID LEARNING VERSUS ITERATIVE LEARNING	43
	2.3.1. Iterative Neuro-computing Learning and Prediction	43
	2.3.2. Rapid Learning as an Alternative	46
	2.3.3. Rapid Learning as a Complement	50
2.4.	DISTINGUISHING RAPID LEARNING BENEFITS	50
	2.4.1. Learning Speed	51
	2.4.2. Network Capacity	53
	2.4.3. Interpretability	54
	2.4.4. User Friendliness	55
	2.4.5. Peak Added Benefit	55
	CONCLUSION	58
	REFERENCES	58

INTRODUCTION

This chapter compares the concurrent learning and information processing system with *selected* pre-programming, iterative, and statistical alternatives. Each selected alternative provides a basis for simple, gross comparison, which is all that space permits in this book. For example, section 2.1 explores only a simple example of pre-programming based on expertise, without considering the powerful available array of related artificial intelligence techniques [1]. Likewise, simple uses of statistical and iterative neuro-computing methods are presented in later sections, without summarizing the many available alternatives [2-6].

Section 2.1 compares the use of rapid learning methods with a pre-programming approach that utilizes programmer expertise. Section 2.2 compares rapid learning methods with a statistical regression approach that utilizes estimation samples. Section 2.3 compares rapid learning methods with an off-line neuro-computing approach. In each of these sections, rapid learning and its alternative are compared and rated on the basis of four criteria: learning speed, network capacity, interpretability, and user friendliness. The final section identifies settings where rapid learning methods are most beneficial.

The three alternative methods described in this chapter provide powerful tools for their designed use — producing input-output functions off-line for later real-time application. Since these alternatives were not specifically designed for rapid learning applications, their low learning speed ratings should come as no surprise. However, other ratings indicate ways that these alternatives can add useful information in on-line learning applications. Still others suggest ways that rapid learning methods may add useful information in off-line applications. The overall conclusion is that off-line analysis methods and rapid learning methods offer several complementary advantages in a variety of applications.

2.1 RAPID LEARNING VERSUS PRE-PROGRAMMING

This section compares rapid learning operation with a standard alternative: pre-programming input-output relationships without performing formal data analysis. The section begins with an industrial process monitoring example, which illustrates how expertise may figure into pre-programmed monitoring functions (see section 2.1.1). The section follows with a description of alternative rapid learning methods (see section 2.1.2). The section ends with a comparison of both approaches that identifies relative advantages of each and highlights their complementary strengths (see section 2.1.3).

2.1.1 Pre-Programmed Prediction

Pre-programmed prediction utilizes user expertise without requiring data analysis. Many monitoring operations are based solely on pre-programmed prediction. In chemical processing plants, for example, operating engineers routinely rely on personal judgments to determine alarm setpoints. In many cases, operators pre-program alarm setpoints by physically positioning alarm switches to suit current plant conditions. In other cases, operators set cutoff-points in computer-based monitoring systems. In all cases, prior *user* training and expertise — rather than prior *data analysis* — governs the nature of prediction functions for monitoring.

Suppose that an engineer is monitoring three redundant tank level gauges in a chemical processing plant (see section 1.1). The engineer may set level alarm cutoff points to satisfy straightforward practical requirements. For instance, warning signals might be appropriate when average gauge values are just below the tank top or just above heating elements near the tank bottom. As part of pre-programming these level cutoff points, the engineer may rely on prior knowledge. For example, the engineer may know from experience that frothy tank liquids require larger safety margins than watery liquids to ensure that overflow or heating element damage will not occur. The engineer may also know from experience what these safety margins should be. As a result, the engineer may be able to set reasonable tank level cutoff points based on prior experience, without requiring specific data analysis results.

When pre-programming is used for forecasting, similar considerations apply, with one exception. Forecast alarms differ from monitoring alarms by predicting eventual rather than imminent danger. For example, forecasts may activate alarms indicating eventual tank overflow or heating element damage. As with monitoring alarms, if pre-programming has been based on level measurements with little variation, then high false alarm rates may result after measurement variances increase. Likewise, if pre-programming has been based on changing conditions, then a low sensitivity to danger may result.

Pre-programming for control typically involves setting parameters for devices called controllers (see section 8.1). Many control applications use PID controllers to maintain settings such as tank levels. PID controllers send control signals that depend on a recent history of previously recorded input signals. Each PID control signal is a weighted sum of three terms: one involving the current signal level, one involving the difference between two recent levels, and one involving the sum of previous values. Ordinarily, PID controllers are pre-programmed by setting the coefficients for the three terms. PID pre-programming may produce effective control systems over long periods, provided that plant conditions do not change significantly. Significant changes, however, may require pre-programming the controllers, an operation that may require significant plant down time to complete.

By using a history of measurements that reflect recent change, PID controllers provide a means for adaptive prediction. Using input variables as signals of change in this way provides a simple and effective basis for reacting to change. Using input variables for adaptive prediction may be effective for monitoring and forecasting as well. For example, plant engineers may have discovered from prior experience that liquid in a tank becomes frothy as temperature increases. Since this implies that alarm cutoff points should have larger safety margins with increasing temperatures, plant engineers can include temperature in alarm setpoint functions accordingly. The simplest way to do this is to include a calibration term in each alarm cutoff value that is a positive weight times a measured temperature value.

Pre-programming may include effective uses of input variables for adaptive prediction, but only up to a point. For example, monitoring two redundant gauges to establish if either gauge is faulty requires establishing a deviance cutoff value. If the difference between any pair of gauge readings exceeds the cutoff value, then an alarm is set. An effective way to establish the alarm cutoff point is to determine baseline gauge variation during normal operation. If difference value cutoff points are set based on resulting variance estimates, alarms should result when and only when gauges are actually faulty. Without using accurate variance estimates, however, arbitrary pre-programmed cutoff values may produce either false alarm rates that are too high or *bona fide* alarm rates that are too low.

Pre-programming without accompanying data analysis thus poses accuracy problems. Along with accuracy limitations, pre-programming includes other problems including the following:

- It requires expert knowledge of potential prediction variables and their relative impact levels.
- It is potentially inaccurate, utilizing only as many input variables as the user can keep in mind.
- It requires substantial user effort to identify and pre-program presumed relationships.
- It is cumbersome to recalibrate if changing plant conditions dictate doing so.

These problems and rapid learning solutions to them will be described more fully in the following section.

2.1.2 Rapid Learning as an Alternative

Rapid learning methods are useful alternatives to pre-programming for many applications. Rapid learning methods avoid awkward pre-programming operations by performing them automatically. For example, redundant gauge monitoring may be implemented automatically by including all redundant gauges in a rapid learning system and predicting each as a function of all others. Rapid learning software can learn how to monitor each gauge in the system effectively after a few measurement trials (see sections 1.1 and 3.2). Other measurements can be included in the system as well to improve monitoring precision, including a recent history of gauge and other measurements such as temperature to achieve a degree of adaptive prediction.

Rapid learning alternatives are especially useful when plant conditions change frequently. For example, suppose that adaptive temperature compensation during monitoring is achieved by including a weight times temperature as a prediction function term (see section 2.1.1). Then *highly* adaptive prediction can be achieved by calibrating the temperature weight concurrently, *during* the continuous monitoring process. Rapid learning software is ideally suited for performing this form of highly adaptive monitoring (see sections 3.2 and 7.1).

Different rapid learning schemes may be used to satisfy different plant operating schedules. One scheme, learning new relationships quickly over a relatively short period, can be applied each time plant conditions change. This scheme, which utilizes equal impact learning (section 3.1.7), is useful in batch processing applications. Suppose that qualitative changes occur occasionally during an industrial process, due to new batches of input chemicals or revised product delivery specifications. At the beginning of each such change, equal impact rapid learning can be initiated (see section 3.1.7 and chapter 5). In this way, new relationships can be learned that do not depend on previous plant operations. A second scheme, using blocked learning (see section 3.1.7), produces prediction functions that depend mostly on more recent measurements, but somewhat on less recent measurements as well. This scheme is more appropriate when gradual changes in plant conditions occur, such as equipment wear and residue buildup. In either case, rapid learning responds to recent changes in measurement relationships that pre-programming may not be able to anticipate.

Concurrent learning and information processing thus offers several benefits in contrast to pre-programming. When monitoring redundant gauges for failure, Rapid learning software (see section 3.1.1) produces the following solutions to the monitoring problems that were identified in the previous section:

- It requires no prior measurement relationship familiarity.
- It utilizes many input measurements at once: input measurements may include functions of all other measurements at each time point, along with recent measurement features.
- It updates learning in real time, compensating for unforeseen effects that would render the prior analysis obsolete.
- It requires no cumbersome periodic recalibration analysis.

	Rapid Learning	Pre-Programmed Prediction
Learning Speed	1	6
Prediction Accuracy	2	2 to 6
Network Size	2	1 to 7
Interpretability	2	3 to 7
User Friendliness	1 to 3	3 to 6

FIGURE 2.1.2 Rating Contrasts Between Rapid Learning and Preprogrammed Prediction (From 1 for "Excellent" to 7 for "Poor" — Courtesy of Rapid Clip Neural Systems, Inc.)

The Figure 2.1.2 ratings contrast concurrent learning methods with pre-programming methods. Like other ratings presented in this chapter, the ratings in the figure are designed to summarize discussion rather than represent empirical findings. Each entry for every category in the figure is a rating from 1 to 7 (1 for "Excellent", 7 for "Poor"), indicating its corresponding method's merit. These are relative ratings that compare pre-programming and rapid learning methods with each other, as well as with iterative and statistical methods that are summarized below (see sections 2.2 and 2.3).

Learning Speed ratings address reaction rate to changing relationships. *Prediction Accuracy* ratings address how closely observed and predicted measurements coincide under normal operating conditions. *Network Size* ratings reflect the number input measurements, output measurements, and measurement features that the method can handle. *Interpretability* addresses how easily the user can determine individual measurement and feature influence on prediction accuracy. *User Friendliness* addresses the overall ease of adaptive monitoring, forecasting, and control implementation.

Rapid learning is assigned the strongest possible rating of 1 for *Learning Speed.* This rating should come as no surprise, given that fast learning is the main design criterion for the rapid learning system. The pre-programmed prediction method receives a much weaker rating of 6. This rating underscores the time-consuming process of subjectively formulating relationships among measurements and programming input-output functions accordingly, whenever plant conditions change.

Rapid learning is assigned a strong *Prediction Accuracy* rating of 2 for two reasons. First, the rapid learning system combines many input measurements and measurement functions to produce optimal prediction functions. Second, rapid learning methods also adjust prediction functions to compensate for changing conditions. Pre-programmed prediction receives a rating range of 2 to 6 for *Prediction Accuracy.* The strong rating of 2 applies in settings that are relatively simple, for example those involving only a few input variables that the user understands fully. The weak rating of 6 applies in settings that are too complicated to be understood fully by the user, especially when relationships are changing over time.

Rapid learning has a strong rating of 2 for *Network Size*, reflecting its capacity to handle many measurements and measurement features relative to statistical and iterative learning alternatives (see sections 2.2 and 2.3). Rapid learning network size capacity is limited, however, because a connection weight is required between each pair of measurement features in the network (see section 4.1 and chapter 5). Pre-programmed prediction *Network Size* is assigned a range of ratings from 1 to 7. The strongest possible rating 1 reflects the capacity of pre-programmed prediction to involve many thousands of inputs when their relationships to outputs are known to the user beforehand. For example, experience may have shown that one million temperature measurements made every minute can be simply averaged to produce a single measurement feature for monitoring other system measurements. Using each of these measurements separately as inputs to a rapid learning system would slow down concurrent temperature sampling considerably, because rapid learning requires updating connections between every pair of the one million measurements.

Although pre-programmed prediction can deal with large network sizes in some applications, effective network sizes are limited in other applications. Typically, the main limiting factor is the user's inability to combine more than a few measurement functions effectively. The weakest rating of 7 for *Network Size* in Figure 2.1.2 reflects this limitation. For example, consider the dilemma of the test engineer who is asked to pre-program prediction functions for strain gauges during aircraft structural testing (see section 1.1). When a test article has been monitored beforehand, relationships among a few strain gauges may be deduced from first-principles analysis prior to the test. However, when a test has never been performed and many strain gauges are involved in the test, relationships among all involved strain gauges are practically impossible to pre-assess. Instead, engineers must rely on measurements that are made during the test to determine prediction relationships. In this and many other settings where the dependency of prediction functions on data increases with network size, pre-programming usefulness is very limited.

The *Interpretability* rating of 2 for concurrent learning reflects the relatively simple structure of concurrent learning prediction functions, along with the broad array of prediction functions that rapid learning software offers. Concurrent learning prediction functions are always weighted sums of measurement features, which may be linear or non-linear functions of input measurements. The weights that link measurement features to prediction functions are estimated and interpreted according to well-established statistical science methods and principles (see chapter 5). Interpretable statistics include standardized weights, correlation coefficients, and partial correlation coefficients, each of which indicates the relative contribution of each input feature, and all of which are available as *Rapid Learner*™ statistics (see chapter 2). In addition, a variety of interpretable, statistically based indices of overall prediction accuracy are provided by rapid learning software.

The *Interpretability* rating for pre-programmed prediction ranges from 3 to 7, indicating fairly strong interpretability for simple networks to weak interpretability for complicated networks. For example, when each of several redundant gauges is predicted simply by averaging all others, the interpretation of each input variable's role is fairly clear. However, as plant conditions and prediction functions become more complex, interpretation becomes far more difficult. For example, when redundant gauges have distinct means and variances, sophisticated interpretations may be required (see chapter 5). Also, when prediction functions combine a variety of input types additively and multipiclatively, prediction roles of individual inputs may be very difficult to interpret.

The *User Friendliness* rating for rapid learning ranges from 1 to 3, indicating straightforward use for some applications, but more difficult use for others. Rapid learning software is user friendly, provided that manual feature function specification and refinement are unnecessary. Rapid learning software has the capacity to process many features quite quickly — enough features at sufficiently high rates that feature refinement may not be necessary in many applications. During monitoring, for example, the software can effectively identify problems like those described in section 1.1, when records containing 1,000 measurements each arrive every few seconds.

When more measurements arrive at higher rates, however, identifying a reduced number of measurements may be required to satisfy real-time learning and information processing requirements. Refinement operations are automated by *Rapid Learner*™ software up to a point (see section 2.6), but moderately difficult refinement operations may be necessary to deal with such rapidly arriving measurement requirements.

The *User Friendliness* rating for pre-programmed prediction ranges from 3 to 6, indicating moderate to high implementation difficulty. The main difficulties arise during prediction function specification, because these functions must be formulated and programmed manually. Other difficulties associated with recalibration and interpretation were described earlier. All such difficulties make pre-programmed prediction tedious in some applications.

2.1.3 Rapid Learning as a Complement

All three alternatives to pre-programming that are described in this chapter, including rapid learning, require a degree of pre-programming. In selecting input and output variables for any forecasting, monitoring, and control alternative, the user plays a major pre-programming role. The user's pre-programming role is also major in terms of the general method selected as well as the specific models and options selected for the method of choice. As a result, any alternative may be viewed as a complement to pre-programming.

The summary ratings in Figure 2.1.2 identify complementary strengths of rapid learning and pre-programming methods. Since the strongest relative strength of concurrent learning is learning speed, concurrent learning complements pre-programming best when prediction conditions are changing rapidly. Often, as in the examples from the previous section, pre-programmed prediction functions may be used as points of departure for concurrent learning methods. Initial prediction functions that are selected for rapid learning operation may include all variables that have been pre-programmed during previous operation, combined and weighted in the same way. From that point on, prediction function weights may be updated continuously. In applications where pre-programming functions include a sufficiently small number of input measurements and measurement features, new input variables and features may be introduced at the time concurrent learning operations begin. Introducing new functionality in this way may improve prediction accuracy substantially. In addition, the broad array of interpretation statistics that rapid learning software provides may complement pre-programmed prediction interpretability substantially (see chapter 3).

Just as rapid learning methods can complement existing pre-programming methods, the opposite may be true as well. Useful pre-programming to improve rapid learning performance may include careful selection of input variables and feature functions for accurate prediction. Input variable and feature selection is especially important in applications where many potential inputs may be selected, but computer speed and memory restrictions limit the number that may be used.

2.2 RAPID LEARNING VERSUS STATISTICAL ANALYSIS

This section compares rapid learning operation with a statistical alternative: identifying input-output relationships by performing off-line statistical data analysis. The section continues with the industrial process monitoring example from the previous section, by first illustrating how statistical methods might be applied to the same monitoring problem (see section 2.2.1). The section follows with a description of alternative approaches to the same problem using rapid learning methods (see section 2.2.2). The section ends with a comparison of both approaches, highlighting relative advantages of each with a view toward identifying complementary strengths (see section 2.2.3).

2.2.1 Statistical Analysis and Prediction

Statistical models are based on assumptions regarding random measurement generation and correlations among measurements [7-8]. From these assumptions, statistical prediction functions and estimates are deduced and employed with optimal precision in mind. One of the fundamental assumptions behind statistical procedures such as regression is that relationships among measurements are stationary over time. If measurements are stationary then concurrent learning may not be necessary because prediction estimates from a random sample will be appropriate for any future use.

Statistical analysis and prediction methods utilize statistical estimates from representative samples to create prediction functions. In a typical application, a statistician first determines the input and output variables that may be required for solving a monitoring, forecasting, or control problem. An experiment or random sampling scheme is then designed, from which several measurements records will be generated, each containing values of all input and output variables of interest. The generated measurements must represent expected measurements during actual operating conditions; otherwise, prediction functions that are determined based on them will not be appropriate.

Once a sample is generated, prediction functions are determined using statistical regression or a similar prediction method [5]. Regression identifies input variables that accurately predict output variable values in the sample. Once input variables have been determined, weights linking each input variable to each output variable are estimated.

Statistical procedures like regression are very powerful, because they offer a large array of variable selection, prediction model, and estimation methods to choose from. Indeed, comprehensive statistical software packages such as SAS™ include so many options that multi-volume technical manuals are provided to describe them [9]. These options are powerful in terms of providing high prediction accuracy and comprehensive interpretability, but substantial statistical expertise is needed to select the best options available for solving a particular problem. In addition, creating statistical prediction models requires substantial prediction statistic interpretation, which is not easy without statistical training and experience.

Trained statisticians can produce valid prediction models from sample data, but practical problem solvers with little statistical training may have considerable difficulty. Subtle problems associated with statistical analysis, such as significance probability interpretation, dealing with small samples can be misleading. Concluding that a prediction function is accurate based on sample results, only to find later that it performs poorly in practice, is an especially common and vexing problem [6].

Just as rapid learning methods can complement pre-programmed prediction, statistical estimation methods do the same. When gauges are monitored for deviant operation, for example (see section 2.1.1), estimates based on representative samples can produce accurate prediction functions. Once these functions are obtained from the sample, alarm setpoints may be used with greater precision than pre-programmed setpoints without statistical analysis.

Pre-programming and off-line data analysis may be appropriate for establishing deviance cutoff values in some settings, but may not suffice in others. For process tank level monitoring, basing cutoff points on a single sample will produce reasonable cutoff values, as long as plant conditions do not change between the time a sample is gathered and the time that gauges are monitored. In many process conditions, however, different chemical batches are introduced over time. Introducing these batches may alter plant conditions significantly. For example, liquid viscosity values may change, significantly altering variances among level gauge difference readings. Likewise, other changing process conditions such as ambient temperature and equipment wear may make baseline values that have been determined from a prior statistical analysis obsolete. Pre-programming may anticipate such conditions up to a point. For instance, providing different prediction functions to deal with different plant conditions may be adequate when the number of possible conditions is small. When plant conditions change frequently and in many ways, however, pre-programmed prediction may be unable to anticipate all changes.

Similar concerns hold when statistical methods are used to complement pre-programming control methods. For example, off-line statistical analysis can provide accurate estimates of PID controller weights from a sample, and these estimates can be utilized effectively if measurement relations have not changed since the time the sample was obtained. When measurement relationships become non-stationary, however, the sample-based controller weights become obsolete.

When contrasted with rapid learning methods in non-stationary settings, statistical methods have several drawbacks. For example, when monitoring prediction functions are produced from statistical regression, the following regression disadvantages result:

- It requires an estimation sample.
- It requires preliminary statistical analysis: the analysis takes considerable time and it must be performed by trained statisticians.
- It requires new samples and analyses whenever relationships among measurements change.

2.2.2 Rapid Learning as an Alternative

Rapid learning methods are useful alternatives to statistical estimation in stationary as well as non-stationary process settings. In stationary settings, rapid learning methods avoid difficult and time-consuming statistical operations by performing them automatically. Avoiding statistical operations not only saves expensive statistician effort but also allows users without statistical training to create prediction models independently. This advantage, which rapid learning shares with off-line neuro-computing (see section 2.3), has generated considerable interest in neuro-computing among engineers and other information processing specialists [2-4].

While rapid learning methods may merely be convenient alternatives in stationary settings, they may be essential in non-stationary settings, especially when measurements are arriving rapidly (see section 2.5). In some non-stationary applications such as commodities forecasting, statisticians may be able keep up with changing relationships on a daily basis (see section 1.2). Results of such daily analyses may be effective when decisions are only made daily. However, when decisions are made and relationships are changing every minute, updating learned relationships using manual statistical analyses is not possible. In these and many other rapidly changing settings, rapid learning methods may be far more powerful.

Concurrent learning and information processing thus offers several benefits in contrast to off-line statistical estimation. When monitoring redundant gauges for failure, rapid learning software (see section 2.1.1) produces the following solutions to the monitoring problems that were identified in the previous section:

- It requires no estimation sample.
- It requires no preliminary statistical analysis.
- It requires no follow-up statistical analysis when measurement relationships change.

The Figure 2.2.2 ratings contrast rapid learning methods with statistical estimation methods. (The ratings in Figure 2.2.2 have the same basis and interpretation as those that were described in section 2.1.2 — the rapid learning ratings in the two figures are identical.) Statistical learning speed is given a broad range of ratings from 3 to 7. The stronger rating of 3 applies in settings like commodities trading that have a degree of routine statistical updating capacity. For example, a trader who is trying to monitor stock prices 30 days into the future might hire a statistician to create prediction functions every day. The statistician may first gather an estimation sample made up of potential independent variables along with stock price values. The test sample is arranged into records, with current values of independent variables, a recent history of their values going back per-

	Rapid Learning	Statistical Prediction
Learning Speed	1	3 to 7
Prediction Accuracy	2	1 to 5
Network Size	2	2 to 6
Intcrprctabilility	2	1 to 7
User Friendliness	1 to 3	4 to 7

FIGURE 2.2.2 Rating Contrasts Between Rapid Learning and Sample-Based Statistical Prediction (From 1 for "Excellent" to 7 for "Poor" — Courtesy of Rapid Clip Neural Systems, Inc.)

haps 60 days, and stock price values to be predicted perhaps 30 days into the future. The statistician then uses statistical procedures to select input variables and input-output functions that forecast prices most accurately in the sample. The statistician then automates the process to some extent, so that at the end of each trading day new input-output prediction weights are estimated from an updated sample. Each updated sample is obtained by inserting measurements for the current day into the sample.

The key concern for statistical estimation learning speed, then, is the degree that sample updating and re-analysis from time to time can be automated. The *Learning Speed* rating range of 3 to 7 in Figure 2.1.2 reflects a wide degree of automating degree. By contrast, the rapid learning rating of 1 indicates a high level of learning automation. This rating should come as no surprise, since rapid learning methods have been developed with automated learning in mind. Indeed, rapid learning methods may be viewed as nothing more than statistical regression methods with completely automated sample and estimation updating (se chapter 5). In terms of stock market forecasting and related applications, such as airline and hotel reservation forecasting, the statistical approach is widely used for daily updates. Hourly updates may also be possible with considerable pre-programming and intense statistician effort. However, rapid learning methods offer the only alternative for much higher learning rates.

Statistical method *Prediction Accuracy* ratings in Figure 2.2.2 range from 1 to 5. The strongest rating in the range applies in stationary settings, where statisticians with sufficient sampling resources and time can identify the best possible input-output functions. By contrast, some prediction function identification tools are not available with rapid learning methods because fast learning requirements bar their use. The weaker rating applies in rapidly changing relationship settings, such as commodities price forecasting every minute (see section 1.2), where statistical methods cannot update learned relationships fast enough to produce accurate predictions.

Statistical method *Network Size* ratings in Figure 2.2.2 range from 2 to 6. The strongest rating is the same as the rapid learning rating because statistical methods, like rapid learning methods, require a weight for linking all input-output measurement features with each other (see chapter 5). The weakest rating reflects a basic limitation in standard statistical methods such as regression: the need to have more records in an estimation sample than the number of input variables and output variables. This limitation, which is not shared by rapid learning methods (see chapter 5), restricts statistical estimation use in many applications such as test, survey, and questionnaire item analysis, where individuals in small samples respond to many items.

Statistical method *Interpretability* ratings in Figure 2.2.2 range from 1 to 7. The strongest rating in the range applies in settings where statisticians with sufficient time can generate highly informative reports, using available statistical tabulating and summarizing tools [6]. By contrast, prediction function identification tools available with rapid learning methods are limited because fast learning requirements bar their use. The weaker rating applies when statistical expertise and time for interpreting statistical results is unavailable. This problem, which occurs often in practice, is vexing for the user who must interpret complex statistical results without the necessary statistical training.

Statistical method *User Friendliness* ratings in Figure 2.2.2 range from 4 to 7. The lowest rating reflects the worst possible case where the untrained statistician must use and interpret statistical methods in complicated situations. The relatively weak rating of 4 reflects the author's experience that in most applications, using statistical methods to identify prediction functions is not easy.

2.2.3 Rapid Learning as a Complement

Rapid learning methods may be viewed as statistical methods that have been selected and modified for rapid learning speed (see chapter 5). As the previous section has described, this selection and modification process has produced methods that are most useful in non-stationary settings. However, trained statisticians may produce more accurate results in stationary settings, and they may produce more interpretable results in general. Statistical and rapid learning methods may thus complement each other in several ways.

In many non-stationary settings, existing statistical results may be used as points of departure for rapid learning methods. Input-output variables and regression weights that have been identified from prior samples may provide initial rapid learning software configurations. During follow-up operations from that initial point, the software can adjust connection weights adaptively for continued accurate prediction.

Statistical interpretation methods may also be used periodically during rapid learning operation. For example, connection weights may be sent by *Rapid Learner*™ software to an output file occasionally, and they can be supplied to statistical software for analysis (see chapter 7). The statistician may then use sophisticated procedures that are not available in *Rapid Learner*™ software, for example step-wise regression and factor analysis methods available in SAS™ [9], to perform highly sophisticated model refinement operations. Likewise, deviance statistics can be supplied to statisticians who may perform sophisticated residual operations [6], which are not possible using *Rapid Learner*™ software.

Although they have not been designed for off-line estimation, rapid learning methods may usefully complement statistical methods off-line, especially when statistician availability is limited. Since rapid learning methods can perform prediction at any point during learning, they can also perform prediction only after learning from an entire estimation sample. As a result, statistical operations such as off-line regression followed by on-line prediction can be performed relatively simply by rapid learning methods.

2.3 RAPID LEARNING VERSUS ITERATIVE LEARNING

This section compares rapid learning operation with an iterative neuro-computing alternative: using off-line training to identify input-output relationships. The section begins with a review of iterative neuro-computing methods, when applied to a phoneme recognition case study (see section 2.3.1). The section follows with a description of alternative approaches that may be taken to the same problem using rapid learning methods, including comparative empirical results from the same case study (see section 2.3.2). The section ends with a comparison of both approaches, highlighting relative advantages of each with a view toward identifying complementary strengths (see section 2.2.3).

2.3.1 Iterative Neuro-computing Learning and Prediction

A rich variety of iterative learning schemes have been developed and published in the academic neural network literature during the last 10 years [2-4]. This section emphasizes the scheme that has received the most attention: using the back-propagation learning algorithm to train multi-layer perceptron models [2]. Multi-layer perceptron models have pre-specified input and output variables, along with an internal linkage structure. In its most common form, the linkage structure includes connections from each input to *hidden layer neurons*. The output of each hidden layer neuron is a pre-specified increasing function of the cumulative input to the neuron. The cumulative input is a weighted sum of input values to the neuron. Each hidden layer neuron, in turn, is connected to output neurons in the same way that inputs are connected to it. As a result, multi-layer perceptron

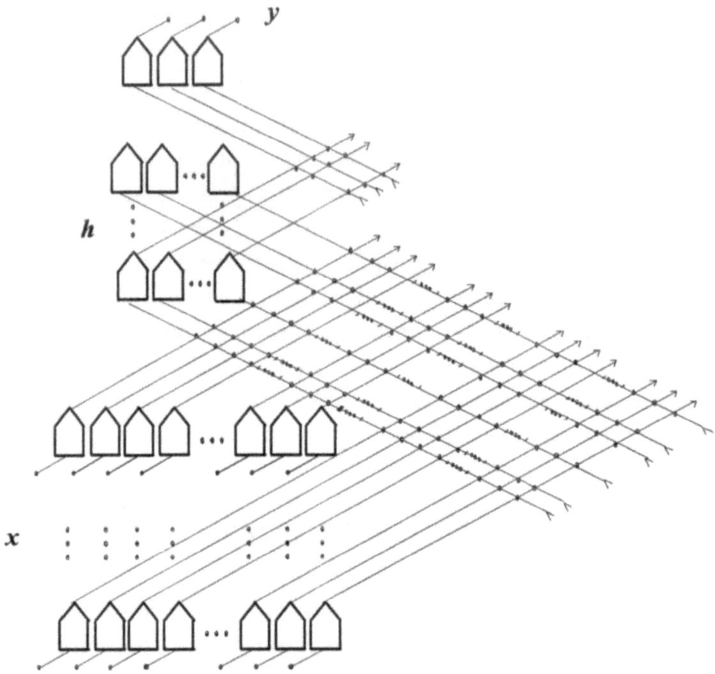

FIGURE 2.3.1 A Multi-layer Perceptron for Phoneme Recognition (Courtesy of Rapid Clip Neural Systems, Inc.)

output values depend on input values along with two sets of weights: one set connecting input values to hidden layer neurons and one set connecting hidden layer neuron output values to output neurons.

Multi-layer perceptron model structure can be tailored to specific applications needs. Figure 2.3.1 shows a model that has been tailored to a phoneme recognition application [10-12]. The input variables represented by the matrix ***x*** have been arranged in rows and columns. Each row of ***x*** corresponds to an audible frequency band and each column corresponds to a time slice during which a phoneme is pronounced. Each time a phoneme is received all measurements in ***x*** are recorded. The input measurements in ***x*** are transformed into several weighted sums, each of which is received by a hidden layer neuron. The hidden layer neurons are shown in Figure 2.3.1 as a matrix ***h***. Each neuron in ***h*** produces an output value that depends on its input weighted sum. The output values from ***h*** are also transformed into several weighted sums, each of which is received by an output neuron. The output neurons are shown in Figure 2.3.1 as a vector ***y***. Each element in ***y*** corresponds to a spoken phoneme, such as "B," "D," or "G". For each set of input values in $\boldsymbol{x}$, the system recognizes whether the input corresponds to a "B," "D," or "G" by identifying which of the three $\boldsymbol{y}$ neurons has the highest output value.

Multi-layer perceptron learning is accomplished off-line, using the backpropagation learning algorithm and data from a training (estimation) sample. Just as statistical regression weights are obtained from an estimation sample, perceptron connection weights are learned from a training sample. In both cases, weight estimation/learning procedures are designed to predict output values that are close to observed output values in samples. In both cases also, estimates are obtained off-line when output values are known, for later on-line use when output values are unknown.

Multi-layer perceptron models differ from statistical regression models in a basic way that has several interesting consequences. Statistical regression models completely pre-specify input-output functions in polynomial terms. That is, once a statistical regression model has been specified prior to estimation, each output can be expressed as a weighted sum of input values or powers and cross-products among input values. Once the statistical model has been pre-specified, the only remaining requirement for complete input-output specification is estimating weights linking each term to each output. This requirement is met during off-line estimation.

In contrast with statistical regression models, multi-layer perceptron models do not pre-specify input-output functions in polynomial terms. Instead, the backpropagation algorithm may be viewed as learning polynomial structures automatically, along with connection weights, during backpropagation training [13]. Backpropagation thus performs learning at two levels, structural model learning and connection weight learning. In contrast, statistical estimation performs regression weight learning only.

Not surprisingly, backpropagation training takes considerably more computer time than statistical estimation because backpropagation produces polynomial models as well as polynomial weights. The main computing requirement for regression weight estimation is covariance matrix inversion (see section 5.1), which takes a relatively small amount of computer time. Backpropagation training, on the other hand, requires many iterations during which estimates are gradually refined. Furthermore, each iteration requires each weight to be adjusted for every record in the training sample. The resulting time for convergence, which may be several hours, may pose significant problems. A related problem is limited network sizes that can be managed by backpropagation training. The number of weights that can be estimated from backpropagation is smaller than the number estimated from regression.

Despite speed and network size limitations, multi-layer perceptron models along with backpropagation training offer a key advantage over statistical estimation: the capacity to learn input-output models automatically, without requiring statistical expertise and effort. This automatic modeling capacity has generated a great deal of recent interest in neural network methods. To illustrate this neuro-computing advantage, a comparison study was performed with phoneme recognition data [13]. The statistical approach utilized a first-order model, in which all inputs in Figure 2.3.1 were linked to each output. It also utilized second-order and third-order models, in which selected cross-products and triple products among the inputs were provided as well. The statistical approach utilized SAS™ step-wise discriminant analysis. An alternative neural network approach was used with the Aspirin-Migraine package [14]. The two approaches yielded comparable correct classification rates, but the statistical approach required much more user effort. Furthermore, the statistical approach required much more user expertise than the neural network approach.

From an applications viewpoint, then, iterative neural network approaches offer the promise of identifying input-output functions from training data off-line, with less user training and effort than statistical approaches. Neural network applications have drawbacks, most notably slow iterative convergence and limited network size capacity. Yet their appeal of making data analysis possible for non-statisticians has created a new generation of users who perform prediction without prior statistical training.

However, if real-time learning in non-stationary settings is required, iterative neural network methods have similar limitations to statistical methods. For example, adaptive learning limitations of backpropagation training include the following:

- It requires a training sample.
- It requires preliminary network configuration and iterative training.
- It requires new sampling and iterative training whenever relationships among measurements change.

2.3.2 Rapid Learning as an Alternative

Like iterative neuro-computing methods, rapid learning systems may be used as prediction tools by users without prior statistical training. Rapid learning methods are most valuable as alternatives in applications for which they were designed: fast learning while relationships among measurements may be changing. Unlike iterative neuro-computing methods, rapid learning methods break learning into two parts (see section 4.1): connection weight learning, which may be accomplished in real time; and model refinement learning, which may be accomplished occasionally or in parallel. This strategy of learning separation permits rapid learning methods to adapt very quickly, without a training set.

When compared to backpropagation training for such applications, rapid learning software offers the following advantages:

- It requires no estimation sample.
- It requires no preliminary setup and iterative training.
- It requires no new sampling and iterative training whenever relationships among measurements change.

Along with being relatively simple, rapid learning produces prediction functions that are relatively accurate. To contrast rapid learning with off-line neurocomputing accuracy, data sets were generated and analyzed for a pattern recognition case study that has already been introduced (see section 1.4). The data sets were generated by producing noisy versions of the Figure 1.4.2 data and by creating a training data set and an independent test data set from the resulting records. (Noisy observations were obtained by switching each pixel value in the figure to its noisy counterpart with a probability of 0.1). Backpropagation training and rapid learning methods were applied to obtain pattern classification functions in the training set. Incorrect classification rates in the test set were computed after estimates were obtained from the training data.

Two alternative concurrent learning prediction models were used. A first-order model predicted output pattern type as a weighted sum of the 49 input pixel values. A second-order model predicted output pattern type as a weighted sum of the 49 input pixel values along with the 1,176 products among each pair of input pixel values. A fully connected, single hidden layer perceptron was also used, with 16 neurons in the hidden layer.

The experiment produced incorrect classification rates in the test set based on all three models, after prediction functions were estimated from training samples of various sizes. For the two rapid learning models, results based on consecutive sample sizes ranging from 1 to 48 were used. For the perceptron model, sample sizes of 4, 8, 16, 32, and 40 were used. The results were produced in this way to indicate how accurately each method would perform at each time point, after learning up to that point.

Incorrect classification rates were obtained in the test set by using each prediction model that had been learned up to each time point. For each record in the test set, predicted probabilities for each of the 16 pattern types were computed as learned functions of the pixel values for that record. Incorrect classification rates were computed as the proportion of records in the entire test set that were incorrectly classified as the wrong type.

Figure 2.3.2.1 shows incorrect classification rates for backpropagation learning, first-order rapid learning, and second-order rapid learning. All three models produced increasing accuracy as a function of the number of learning trials, as expected. All three models also produced comparable accuracy levels

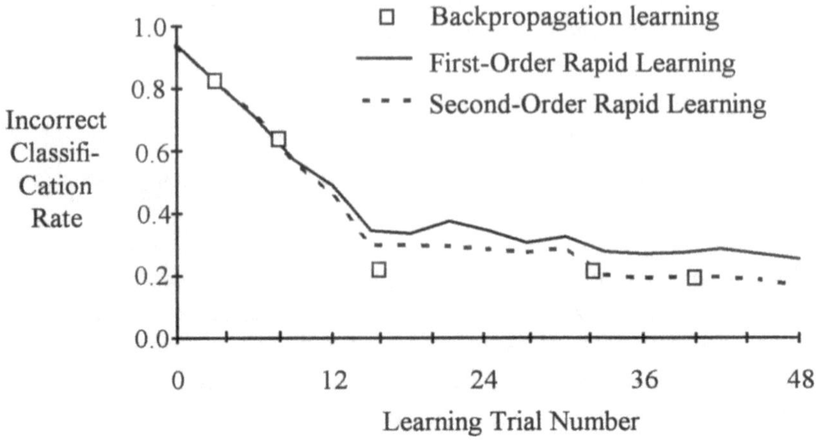

FIGURE 2.3.2.1 Pattern Classification Comparison Results
(Courtesy of Rapid Clip Neural Systems, Inc.)

from trial 1 to 8, indicating that comparable performance levels could be expected during early learning. The second-order rapid learning model and the backpropagation model performed slightly better beyond trial 8. This was also expected, because of the underlying second-order relationship among the pixels and pattern types (see section 1.5).

Several interesting conclusions follow from the Figure 2.3.2.1 results. First, rapid learning, as well as backpropagation learning methods, may produce similar accuracy levels, in the presence of non-linear relationships. Second, the two methods can be used to update learning from trial to trial, if time permits doing so. Third, since rapid learning methods are fast and fully automatic, they can learn adpatively after a few learning trials, even when measurement records arrive very quickly. For this particular problem, concurrent learning can keep up with records arriving every second or less (see section 2.4.1). Finally, since off-line neuro-computing methods require a training set, manual setup time, and iterative training, they cannot adapt nearly as quickly — for this particular problem, the five backpropagation results in Figure 2.3.2.1 required many hours of user setup time and iterative training.

The Figure 2.3.2.1 results also illustrate how quickly concurrent learning methods can adapt to changing conditions. The results in the figure were obtained from the first trial onward in a training sample, but similar results could prevail immediately after any point in a changing process. For example, if changing input patterns created new second-order relationships between pixels and pattern types, concurrent learning methods could adapt to the new relationships just as quickly. Finally, the results in the figure show that off-line neuro-computing methods can adapt as accurately as rapid learning methods in principle. How-

	Rapid Learning	Iterative Learning
Learning Speed	1	3 to 5
Prediction Accuracy	2	2 to 6
Network Size	2	6
Interpretabilility	2	4 to 7
User Friendliness	1 to 3	2 to 6

FIGURE 2.3.2.2 Rating Contrasts Between Rapid Learning and Iterative Neuro-Computing Prediction (From 1 for "Excellent" to 7 for "Poor" — Courtesy of Rapid Clip Neural Systems, Inc.)

ever, they cannot adapt nearly as quickly in practice — iterative neuro-computing learning rates are retarded because their manual convergence time requirements are orders of magnitude slower.

The Figure 2.3.2.2 ratings contrast rapid learning methods with off-line neuro-computing methods. (The ratings in Figure 2.3.2 have the same basis and interpretation as those that were described in section 2.1.2 — the rapid learning ratings in the two figures are identical.) The learning speed ratings for iterative learning are higher than their statistical estimation counterparts (see Figure 2.2.2) because reduced manual effort prevails over slow convergence time in the author's judgment. Prediction accuracy is given a range of ratings. The strong rating of 2 reflects comparable accuracy to rapid learning in settings like the Figure 2.3.2.1 case. The weaker rating of 6 reflects two problems: difficulties associated with adapting rapidly to changing conditions in practice, and certain convergence problems associated with the backpropagation method and similar iterative methods [13].

The *Network Size* rating of 6 reflects low model complexity capacity relative to rapid learning methods. For example, the second-order results in Figure 2.3.2.1 were obtained from a rapid learning model that included 49 individual input variables and 1,176 cross-products among the individual input variables. Rapid learning software can receive each of these measurements, update learning, and predict *each* input as a function of *all* others in about 1 second. Alternative software using the backpropagation algorithm has far less network size capacity.

The lowest ratings for iterative learning are assigned in the *Interpretability* category. Rapid learning shares many useful interpretation features with its close relative, statistical regression (see chapter 5). Such features include soundly based indices of prediction accuracy, individual input contributions to overall prediction accuracy, and step-wise contributions to overall prediction accuracy These features result because in both cases, (a) outputs are expressed as weighted sums of input values and products among input variables; and (b) the underlying models are *identifiable* (see section 4.6). In contrast, iterative learning methods have no such clearly defined and soundly based interpretability statistics. Instead, they must rely on less formal performance indices, which are far more difficult to interpret [13].

The *User Friendliness* ratings for iterative learning methods range from better than rapid learning ratings to worse. The stronger rating of 2 reflects rare settings where third-order or higher-order relationships may exist among measurements. Identifying such relationships may require manual effort when rapid learning methods are used, especially if many input variables are involved. In such cases, iterative learning may produce more accurate input-output relationships with less user effort. The weaker rating of 6 reflects the need in more common settings for manual user effort, especially gathering a representative training sample each time relationships among measurements change.

2.3.3 Rapid Learning as a Complement

The Figure 2.3.2.1 ratings indicate complementary uses of iterative learning and rapid learning methods. Iterative learning methods may be no worse in stationary settings where a single training effort may be used indefinitely. If rapid learning methods are used routinely in such settings, iterative learning methods may be used to augment performance when highly non-linear input-output relationships are suspected.

If iterative learning methods are used routinely in stationary settings, rapid learning methods may be used as well to provide interpretability statistics. End users who would like to know if certain inputs influence output prediction more than others may obtain complementary results from rapid learning software that will supply such information.

If iterative learning methods are used routinely in slowly changing settings such as daily stock forecasting (see sections 1.2 and 2.2), rapid learning methods may produce results that will improve iterative learning performance. For example, rapid refinement methods have the capacity to combine redundant inputs and remove unnecessary inputs automatically. Since the number of inputs is a basic iterative learning limitation, complementary rapid refinement results may be obtained occasionally to increase overall prediction effectiveness.

2.4 DISTINGUISHING CONCURRENT LEARNING BENEFITS

This section highlights key benefits of rapid learning methods. The section begins by highlighting the key distinguishing benefit: the capacity of rapid learning methods to learn at very high record arrival rates (see section 2.4.1). The section follows by comparing network size capacities (see section 2.4.2). The section then deals with less tangible, but equally important, rapid learning benefits associated with interpretability (see section 2.4.3), as well as user friendliness (see section 2.4.4). The section ends by identifying application profiles where rapid learning methods achieve their highest relative value.

2.4.1 Learning Speed

From examples and results that have been presented so far in this book, the critical role of learning speed for concurrent learning and information operations should now be clear. The results presented in chapter 1 either involve hundreds of variables arriving frequently, or they can easily be extended to do so. When relationships among hundreds of variables must be learned, connection weights numbering in the tens of thousands must be estimated. For example, 10 measurements have 45 pair-wise connections, 100 have 4,950 connections, and 1,000 have 499,500 interconnections. Statistical estimation and iterative learning methods are not suitable for rapid learning because they have not been designed to update connections quickly. For example, statistical regression requires computing and inverting a sample covariance matrix during prediction weight estimation, and backpropagation learning requires many passes through the entire training sample.

Rapid learning methods provide the following benefits for rapid connection weight updating:

- They do not require matrix inversion.
- They do not require iterative operation.
- They may employ massively parallel processing on digital chips.
- They may employ massively parallel processing on analog chips.

When implemented on a conventional computer having only one central processor, the single processor must update all connection weights among all measurements or measurement features (see section 3.1 — for purposes of this discussion, "features" is synonymous with "measurements"). Since the number of connections increases with the number of features squared, conventional computer learning rates decrease with the number of features squared. Relatively slow learning results when only conventional processors are used.

When parallel processors are used, much higher learning can be achieved. Very fast parallel chips have been specifically designed for connection weight updating (see section 2.1): a digital version has one digital processor per feature and an analog version has one analog processor dedicated to each required arithmetic operation. Since the digital version has one processor per feature instead of one processor overall, digital chip processing speed decreases only with the number of features instead of with the number of features squared. Moreover, since the analog version has one processor for each required arithmetic operation, processing speed does not decrease with the number of features at all. The overall result is digital chip learning that is much faster than conventional computer learning, and analog chip learning that is much faster than even digital chip learning.

Figure 2.4.1 lists rapid learning response times that may be achieved using alternative processors. The left column lists the number of measurement features to be interconnected. The second column shows that updating times using a conventional computer increase exponentially with the number of features, as expected. The third column shows that updating times using a digital parallel chip increase linearly with the number of features. The right column shows that updating times using a parallel analog chip do not increase with the number of features. Some entries are blank in the figure because current technology limits the number of features and interconnections that may reside on a single chip.

During each time value listed in Figure 2.4.1, the following operations are performed by the rapid learning Kernel module (see section 4.1):

- It predicts each feature as a weighted sum of all other features.
- It predicts all missing features as weighted sums of all non-missing features.
- It updates feature connection weight learning.
- It updates other learned statistics necessary for concurrent learning and prediction.

Fea-Tures	Conventional Computer* (milliseconds)	Digital Chip† (microseconds)	Analog Chip†† (nanoseconds)
64	7	60	50
128	26	120	50
256	99	240	50
1,024	2,088	—	50
2,048	7,084	—	—

FIGURE 2.4.1 Representative Rapid Learning Response Times (*Obtained from a Pentium™ 100 megahertz computer with 16 mega-bytes of memory and running Windows 95™; †Deduced from a parallel digital chip model, assuming 100 megahertz clock rates; ††Deduced from an analog chip model, assuming 3 nanosecond delays for each arithmetic operation — Courtesy of Rapid Clip Neural Systems, Inc.)

Once these operations are performed by the Kernel module, most remaining operations for concurrent monitoring, forecasting, and control require relatively little time. The response times in the figure, then, may be viewed as benchmarks for ideal concurrent operation.

The conventional computer response times in Figure 2.4.1 are fast enough for most applications that are described in chapter 1. For example, kernel operations for predicting each of 256 strain gauges from all others require 25 milliseconds per record. As a result, rapid learning operation can keep up with strain gauges arriving every second, even if input, output, and other operations take 10 times as long. Likewise, conventional computers can keep up with many other rapid learning operations, when a small number of features must be processed several times per second and when a larger number of features must be processed several times per minute.

The digital and analog chip response times in Figure 2.4.1 are fast enough for a variety of high-speed operations, ranging from Internet search monitoring to computer memory access. When only speed is considered, analog chips have a clear advantage. Digital chips have potential accuracy advantages, however, because digital accuracy can be governed precisely while analog accuracy depends on variable chip material properties. Despite these accuracy concerns, the chip response times in the figure show that rapid learning technology is feasible for a variety of very high-speed monitoring, forecasting, and control applications.

It should be noted that the conventional computer response times in Figure 2.4.1 are obtained from a personal computer running on Windows 95, rather than on a more powerful machine and operating system. Other benchmarks have shown that more powerful machines can operate much more quickly and can handle many more features per record. For example, benchmarks were obtained from a Silicon Graphics Indigo computer (running MIPS 4400 at 250 megahertz, with 64 megabytes of random access memory), showing Kernel operation rates of 0.28 seconds per 1,024-feature record, 1.22 seconds per 2,048-feature record, and 6.1 seconds per 4,096-feature record.

2.4.2 Network Capacity

As the feature size values listed in Figure 2.4.1 show, rapid learning operations can handle many features using conventional computer technology. Feature size is ultimately limited only by available memory and required speed. The number of computer memory bytes required for concurrent operation is proportional to the number of features squared. If a conventional computer has a large amount of random access memory, then concurrent operations involving many features may proceed without time-consuming swaps into and out of disc memory. Otherwise, swapping delays result. If swapping delays are acceptable, modern computers containing multiple giga-byte disks can handle tens of thousands of features. Even without swapping delays, however, response times decrease substantially as feature sizes increase. For example, the Figure 2.4.1 conventional computer time values extrapolated to 32,000 features indicate a response time of over 4 minutes per record. This delay will be too long to meet some concurrent learning and information processing requirements.

Even when diminishing speed returns with increasing feature sizes are taken into account, rapid learning methods offer a substantial feature size advantage, relative to iterative learning methods. Iterative learning technology has roughly the same tradeoff between memory and connections as rapid learning technology, but it has the added disadvantage of requiring increasing numbers of iterations for convergence as the number of features increases. This disadvantage effectively reduces network size, while rapid learning, being non-iterative, has no corresponding disadvantage. The rapid learning network size advantage is greater still in monitoring applications, because rapid learning networks routinely predict each measurement as a function of all others. By contrast, multi-layer perceptrons do not treat variables as inputs as well as outputs within the same network. As a result, several networks are needed to monitor each variable as a function of all others.

As the blank entries in Figure 2.4.1 suggest, concurrent learning network capacity for chips is more limited than for conventional computers. This limitation occurs because the chip speed advantage exists only if all required connection weights and processors for rapid learning reside on the same chip. Otherwise the speed advantage is lost, because current chip technology does not provide the necessary communication capacity between chips for rapid connection weight learning. However, emerging multi-chip technology offers the promise of solving the inter-chip communication problem. Once that problem is solved, rapid learning feature sizes much larger than those shown in Figure 2.4.1 will be possible.

Despite network size limitations, the feature set sizes indicated in Figure 2.4.1 can solve many rapid learning problems. In many high-speed applications, predicting a measurement as a function of 255 others can be highly precise. In other very high-speed applications that require smaller feature set sizes, rapid refinement methods may be used to reduce the number of necessary features (see sections 1.5 and 8.1). In any event, rapid learning has a decided advantage for such high-speed operations because comparable statistical or iterative operation within the Figure 2.4.1 limitations is not feasible.

2.4.3 Interpretability

Rapid learning models and results are easy to interpret because they are simple; they *must* be simple in order to be so fast. They are also easy to interpret because they are based on a sound statistical model (see section 4.4), for which a variety of interpretation statistics has been developed. These *Rapid Learner™* software statistics, which describe prediction accuracy (see section 2.2), are explained in terms that do not require prior statistical training. For users with more sophisticated model interpretation needs, supplementary interpretation statistics are provided as well (see chapter 7).

2.4.4 User Friendliness

Rapid learning models are user friendly for several reasons, most importantly because they are designed for automatic adaptation without training data. By removing the training data requirement, rapid learning methods allow users to begin concurrent learning and performance operations at once, without user effort. Likewise, rapid learning methods remove the need for gathering and analyzing new samples when plant conditions change.

Rapid learning methods are also user friendly because they have the capacity to process many prediction functions involving many independent variables at once. By providing automatic prediction of all measurements from all others, rapid learning methods avoid time-consuming model-building for each measurement individually. Also, since rapid learning methods may use thousands of independent variables for prediction, they may include many polynomial and historical measurement features in prediction functions automatically, without requiring time-consuming variable selection operations by the user.

In summary, rapid learning methods are user friendly for the following reasons:

- They do not require manual training sample gathering.
- They do not require manual training sample analysis.
- They do not require manual re-training when process conditions change.
- They do not require manual refinement operations.

2.4.5 Peak Added Benefit

Rapid learning provides peak value when measurement relationships must be learned rapidly and prediction accuracy is critical. Peak added benefits that are illustrated by the examples in this book include the following:

- Providing early warnings of strength changes during original structural testing.
- Providing immediate warnings of gauge malfunction during changing plant conditions.
- Providing valuable price forecasts under volatile trading conditions.

- Providing missile tracking control signals during unexpected excursions.

When relationships change, rapid learning utility increases with increasing prediction frequency. For example, suppose that trade decisions are based on price forecasting (see section 1.2). Suppose further that the same profit margins can be expected for decisions made every minute as for decisions made every day. Under these conditions, rapid learning methods add value as a function of increasing prediction frequency, because more profitable decisions can be made per week on a minute-by-minute basis than on a day-by-day basis. Similar concerns hold for process monitoring and control decisions that are indirectly tied to profitability. Expensive down time can be avoided by early warning of impending failure — the earlier the better. Likewise, precise control can be achieved by rapidly adaptive control, the faster the better.

Increasing rapid learning utility with increasing prediction frequency has special importance in these times of rapid information processing growth. Increasingly, monitoring, forecasting, and control decisions are being made based on electronic information, arriving at astounding rates that are growing from year to year. Rapidly adaptive prediction has the potential to add value at the same rates. For example, effective monitoring to identify the rogue commodities trader or the fraudulent credit card user at once has the potential to produce big savings. Similar conclusions apply to electronic decisions that are not directly tied to money. Highly adaptive Internet search and computer memory access can produce huge savings in required time and equipment.

Rapid learning methods may or may not add value in settings where relationships are unchanging over time. As the discussion in this chapter implies, rapid learning methods may complement alternative methods in such settings, but alternative methods may be essentially satisfactory. Likewise, rapid learning methods may not add substantial value in settings where relationships change slowly.

Figure 2.4.5 illustrates peak added value for rapid learning methods. Like the ratings presented earlier in this chapter, the plots in the figure are designed to summarize discussion rather than represent empirical findings. The figure plots rapid learning cost and benefit as well sample-based cost and benefit as a function of processing interval. The plots only apply in settings where relationships among measurements are changing and value increases with increasing decision rates.

The Sample-Based Cost plot in Figure 2.4.5 illustrates applications such as stock price forecasting, where periodic sampling and training are required to identify changing relationships. The plot shows that higher cost results when updating must be made hourly instead of daily, because more intense effort is required. At faster intervals, the plot shows sample-based cost leveling off, because no added manual effort can keep up with changes beyond that point. The plot shows sample-based benefit increasing up to that point, but decreasing

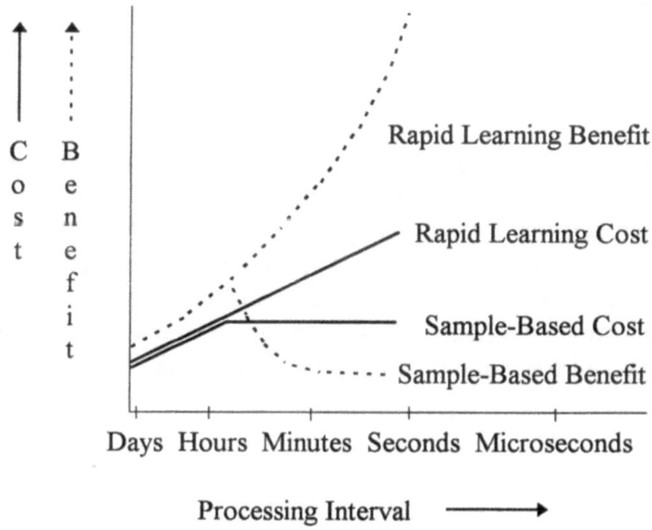

FIGURE 2.4.5 Costs and Benefits Versus Decision Frequency (Courtesy of Rapid Clip Neural Systems, Inc.)

beyond that point, because changing relationships that can no longer be managed beyond that point cause decreased prediction precision.

The Rapid Learning Cost plot in Figure 2.4.5 illustrates increasing cost as processing interval decreases. Reasons for such cost increases include higher computing speed requirements and faster data acquisition needs. The Rapid Learning Benefit plot increases as processing interval decreases unlike its sample-based counterpart, because rapid learning methods can keep up with data arrival rates over the entire processing interval span.

The divergence between Sample-Based Benefit and Rapid Learning Benefit in Figure 2.4.5 illustrates applications where rapid learning has peak added value. As processing speed requirements increase exponentially in changing relationship settings, rapid learning benefits increase exponentially, in sharp contrast with sample-based learning benefits. Peak added benefit for rapid learning methods therefore occurs under the following conditions:

- Prediction accuracy is critical.
- Measurement relationships change over time.
- Benefit increases as a function of information processing speed.
- Information arrival rate is very high.

CONCLUSION

Future Directions

Among several contrasts between rapid learning and alternative methods presented in this chapter, some are quantitative and empirically based, but many are qualitative and judgment based. The qualitative contrasts and conclusions will be strengthened once they have been backed by empirical comparisons. From an applications viewpoint, the strongest conclusions will result from contests comparing rapid learning methods with alternatives. The author welcomes any suggestions, data, and collaboration efforts toward that end.

Summary

This chapter contrasts rapid learning methods with sample-based alternatives in settings where relationships are changing rapidly and accurate prediction is essential. When contrasted with pre-programmed prediction, rapid learning methods provide improved accuracy advantages. When compared with statistical prediction, rapid learning methods offer user-friendliness advantages. When compared with iterative neuro-computing, rapid learning methods offer interpretability advantages. When compared with all three alternatives, rapid learning offers the key advantage of fast, simple, and automatic learning. Some comparisons indicate ways that sample-based alternatives can complement rapid learning methods in on-line learning applications. Other comparisons suggest ways that rapid learning methods may complement sample-based methods in off-line applications. The overall conclusions are that rapid learning methods are ideally suited for changing relationship applications, but sample-based and rapid learning methods offer complementary advantages in other applications.

REFERENCES

1. J. Pearl, *Probabilistic Reasoning in Intelligent Systems: Networks of Plausible Inference*, Morgan Kaufman, San Mateo, CA, 1988.
2. K.J. Aström & T.J. McAvoy, "Intelligent Control: an Overview and Evaluation," in D.A. White & D.A. Sofge (Eds.), *Handbook of Intelligent Control*, Van Nostrand Reinhold, New York, 1992.
3. B. Kosko, *Neural Networks and Fuzzy Systems*, Prentice Hall, Englewood Cliffs, NJ, 1992.
4. D.E. Rumelhert & W.L. McClelland (Eds.), *Parallel Distributed Processing, Explorations in the Microstructure of cognition*, Vol. 1, MIT Press, Cambridge, MA, 1986.

5. N.R. Draper & H. Smith, *Applied Regression Analysis*, New York, Wiley, 1966.
6. F. Mosteller & J.W. Tukey, *Data Analysis and Regression*, Addison-Wesley, Reading, MA, 1977.
7. J. Neyman, "Frequentist Probability and Frequentist Statistics," *Synthese*, Vol. 36, 97-131, 1977.
8. P. Bickel & K. Doksum, *Mathematical Statistics: Basic Ideas and selected Topics*, Holden-Day, San Francisco, 1977.
9. SAS Institute, Inc., *SAS Procedures Guide, Version 6.03*, 4th Edn., SAS Institute, Cary, NC, 1988.
10. A. Waibel, "Modular Construction of Time-Delay Neural Networks for Speech Recognition," *Neural Computation,* Vol. 1, pp. 39-46, 1989.
11. A. Waibel, H. Sawai, & K. Shikano, "Modulatity and Scaling in Large Phonemic Neural Networks," *IEEE Transactions on Acoustics, Speech, and Signal Processing,* Vol. 37, pp. 1888-1897, 1989.
12. A. Waibel, T.K. Hanazawa, G. Hinton, K. Shikano, & K. Lang, "Phoneme Recognition: Neural Networks versus Hidden Markov Models," *Proceedings of the ICASSP International Conference on Acoustics, Speech, and Signal Processing,* New York, 1988.
13. G. Tatman, R.J. Jannarone, & C.M. Amick, "Neural Networks for Speech Recognition: Contrasts Between a Traditional and a Parametric Approach," Unpublished Technical Report, Machine Cognition Laboratory, University of South Carolina, 1994.
14. R. Leighton & A. Wieland, *Aspirin/MIGRAINES User's Manual, Release V4.0*, the MITRE Corporation, McLean, VA, 1991.

[illegible] *Computer Assisted Accounting Analysis*, [illegible]

[illegible]

[illegible]

[illegible]

[illegible]

Part 2

Rapid Learning Features

Part 2 moves from the rapid learning benefits described in Part 1 to the underlying features that produce them (not to be confused with features that are functions of measurements — see Glossary). Chapter 3 introduces rapid learning neuro-computing system features, emphasizing operational features of *Rapid Learner*™ software. Chapter 4 introduces rapid learning models, beginning with neuro-computing models and following with closely related scientific models. Part 2 should give readers a general understanding of the software that produces rapid learning results, along with the underlying models that inspire them.

3

The Rapid Learning Neuro-Computing System

INTRODUCTION . . . 64
3.1. GENERAL SYSTEM OPERATION . . . 64
3.1.1. Real-Time Operation . . . 65
3.1.2. Real-Time, Multiple Processor Systems . . . 66
3.1.3. Simulation Systems . . . 67
3.1.4. Graphical User Interface . . . 68
3.1.5. Operation Specification Overview . . . 73
3.1.6. Data Input Specifications . . . 73
3.1.7. Learning Specifications . . . 77
3.2. MONITORING OPERATION . . . 79
3.2.1. A Strain Gauge Monitoring Example . . . 80
3.2.2. Monitoring Options . . . 80
3.3. FORECASTING OPERATION . . . 81
3.3.1. A Commodities Forecasting Example . . . 81
3.3.2. Forecasting Options . . . 82
3.4. CONTROL OPERATION . . . 85
3.4.1. A Missile Tracking Control Example . . . 86
3.4.2. Control Options . . . 86
3.5. REFINEMENT OPERATION . . . 87
3.5.1. Refinement Options . . . 88
3.5.2. Refinement Statistics . . . 91
3.6. FEATURE FUNCTION PROCESSING OPERATION . . . 93
3.6.1. Feature Function Overview . . . 93

3.6.2. Feature Function Processing Options 94
CONCLUSION 96
REFERENCE 97

INTRODUCTION

Chapter 3 describes rapid learning neuro-computing system operation. The beginning of section 3.1 introduces three system types and outlines how users interact with them. The remainder of the chapter introduces the operational *Rapid Learner*™ software package and describes it in detail. The use of this software package is emphasized, because it is the only available system for concurrent learning and information processing at this time.

Section 3.2 describes monitoring options such as deviance cutoff values for monitoring alarms. Section 3.3 describes forecasting options such as the number of future time points to be forecast. Section 3.4 describes control options such as output control information to be supplied. Section 3.5 describes refinement options such as criteria for combining redundant features into mean features. Section 3.6 describes options for specifying feature functions such as historical features and power features.

Chapter 3 should give readers a general understanding of rapid learning neuro-computing system operation. Chapter 3 should also give enough specific details to prepare readers for *Rapid Learner*™ software use. Comprehensive details are provided in the *Rapid Learner*™ User Manual [1].

3.1 GENERAL SYSTEM OPERATION

This section introduces general characteristics of the rapid learning system including operation on conventional computers (see section 3.1.1), operation with multiple processors (see section 3.1.2), and simulation operation (see section 3.1.3). The section then covers general characteristics of *Rapid Learner*™ software, including its graphical user interface (see section 3.1.4), operational specification types (see section 3.1.5), input measurement and other input specifications (see section 3.1.6), and specifications associated with a variety of learning schemes (see section 3.1.7).

3.1.1 Real-Time Operation

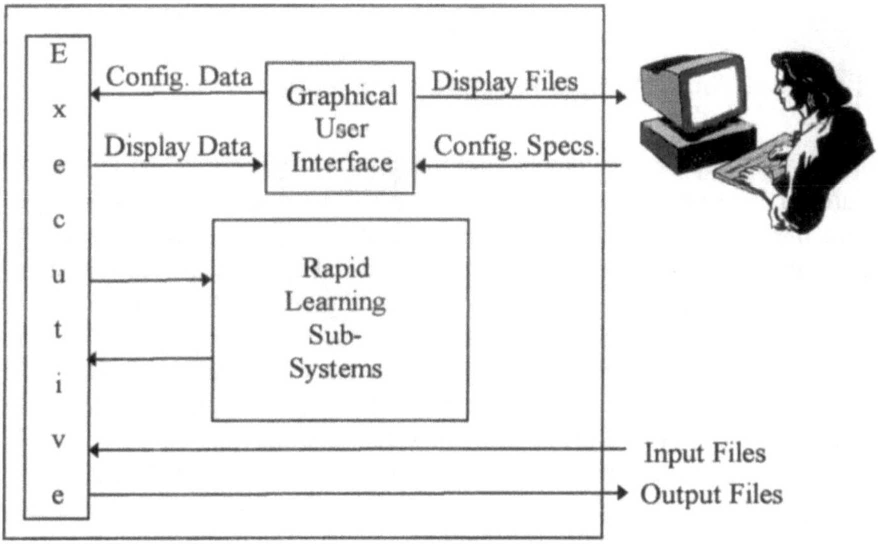

FIGURE 3.1.1 Rapid Learning System Operation Overview
(Courtesy of Rapid Clip Neural Systems, Inc.)

Figure 3.1.1 outlines rapid learning system operation (see section 4.1 and later chapters for internal operation details). System inputs include configuration specifications and measurements from input files. Configuration specifications include input measurement descriptions along with monitoring, forecasting, and control commands. These specifications are supplied prior to initial operation.

The Graphical User Interface in Figure 3.1.1 converts user-supplied specifications to initialization and control commands for rapid learning subsystems (see chapter 4). Once the system has been initialized, concurrent learning and information processing operation begins. Concurrent operation includes reading measurement records and supplying display information continuously. During concurrent operation, display information is converted to graphical display form for user viewing.

At any point, the user may interrupt concurrent operation. For example, the user may decide to change monitoring specifications after noticing recent trends in graphical displays. At any such point, the user may supply modified configuration specifications to the system by reviewing each of the *Rapid Learner*™ specification dialog boxes (see below) and revising them as appropriate.

The line labeled "Output Files" in Figure 3.1.1 indicates a path for optional output information other than graphical display. Such information may or may not be provided by the system, depending on the type of rapid learning operations being performed. Ordinarily, monitoring and forecasting data are not sent to output files. During control operation, however, forecast values and related data are sent to output files for external process control (see section 3.4).

3.1.2 Real-Time, Multiple Processor Systems

Rapid learning operations may be performed on a single computer, as shown in Figure 3.1.1, or on multiple computers. Figure 3.1.2.1 shows one dual computer layout that might be used for process monitoring. The data acquisition device represents a process controller or other device that supplies measurements at regular intervals. The server computer in Figure 3.1.2.1 performs routine mathematical computing and coordinates input and output operations. The client computer coordinates and supplies graphical display information. The system operator supplies system configuration specifications and monitors system operation, but does not monitor process measurements. Instead, a process monitoring specialist uses a different computer for graphical process monitoring.
The Figure 3.1.2.1 dual processor system can easily be expanded to a multiple computer, local area network, client-server system. The multiple server computers can be distributed throughout a process plant, with each client providing distinct graphical monitoring information for a different process specialist. Specialists can have limited control over the information being displayed on their client computers, such as particular gauge selection. However, overall system operation will be controlled by the system operator at the server computer.

Figure 3.1.2.2 shows a multiple computer, client-server system that might be used on the Internet. The system is suitable for transmitting and receiving infor-

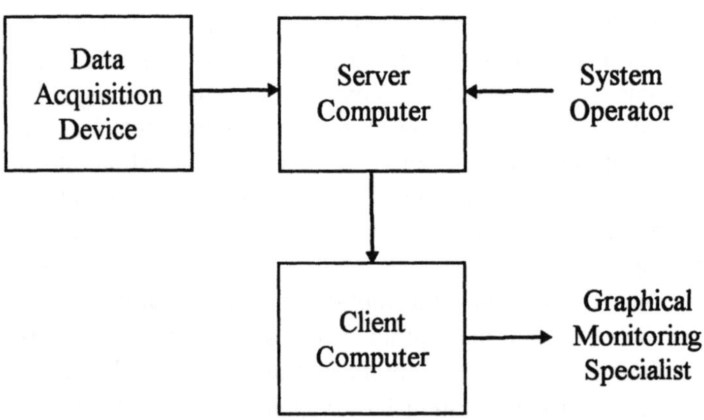

FIGURE 3.1.2.1 A Real-Time, Dual Computer System
(Courtesy of Rapid Clip Neural Systems, Inc.)

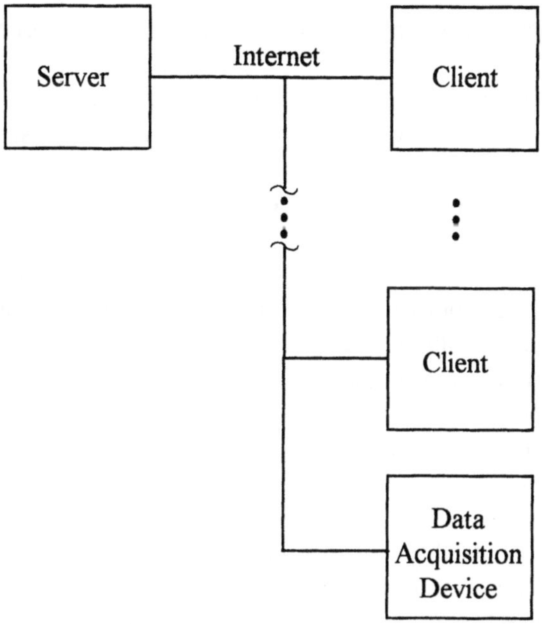

FIGURE 3.1.2.2 A Multi-Computer Client-Server System
(Courtesy of Rapid Clip Neural Systems, Inc.)

mation concurrently, provided that real-time speed requirements are met. Otherwise, transmission delays could slow down system operation considerably. Graphical display transmission delay is a special concern, because graphical files are much larger than measurement files. For applications where remote client-server computers are necessary and transmission delays are significant, restricted networks may be used in place of the Internet. However, Internet-based rapid learning systems remain useful in slower concurrent applications and in batch simulation applications.

3.1.3 Simulation Systems

Pending resource availability (see Preface), *Rapid Learner*™ simulation software is being prepared for Internet use. The simulator software will demonstrate how *Rapid Learner*™ software can solve concurrent information problems, without requiring expensive real-time implementation. Internet users will first download client software to their computers. The software will then be usable for batch simulation operation. Batch simulation includes the following steps:

- The user initiates simulation setup operation on a client computer.

- The user creates and attaches configuration specifications to a client output file.
- The user attaches input measurement data to the client output file.
- The user sends the client output file to a public access server over the Internet.
- The server sets up concurrent operation according to the client output file specifications.
- The server processes the input measurement records in the client output file and creates output display data, as if the measurement records were arriving sequentially, in real time.
- The server sends an output display data file to the client computer over the Internet.
- The user initiates display operation on the client computer, using the display data file.

Pending resource availability (see Preface), specification files and input measurement data for all the examples in this book will be placed into a public-access simulation data directory. These specification files will allow users to perform a wide variety of concurrent learning simulations.

3.1.4 Graphical User Interface

Figure 3.1.4.1 shows a *Rapid Learner™* graphical display for commodities index forecasting at a given point in time (see section 1.2.1). The display shows predicted values at the current minute (39), one minute into the future (40) and five minutes into the future (45). The display also shows a recent history of index values starting at 0. In addition, the display shows the values that were predicted at a previous point in time (29), along with the values that were forecast at that point in time, one minute into the future (30), and six minutes into the future (35).

Sets of predicted values and their tolerance bands, like the two sets in Figure 3.1.4.1, are called <u>telescopes</u>. The term describes their visionary role in that they look into the future. The term also describes their boundaries that become wider as they move into the future. Increasing widths as tolerance bands project into the future reflect increased uncertainty as the forecasting time span increases.

Graphs like Figure 3.1.4.1 are routinely generated for forecasting and/or monitoring by *Rapid Learner™* software. These graphs always include the current measurement value and its tolerance band, along with a recent history of measurement values. For forecasting applications, the graphs also include telescopes. The graphs may also include previously predicted values like the thick line beginning at minute 29. This line, which shows what the forecast values were at minute 29, can be compared visually to the actual values that have been

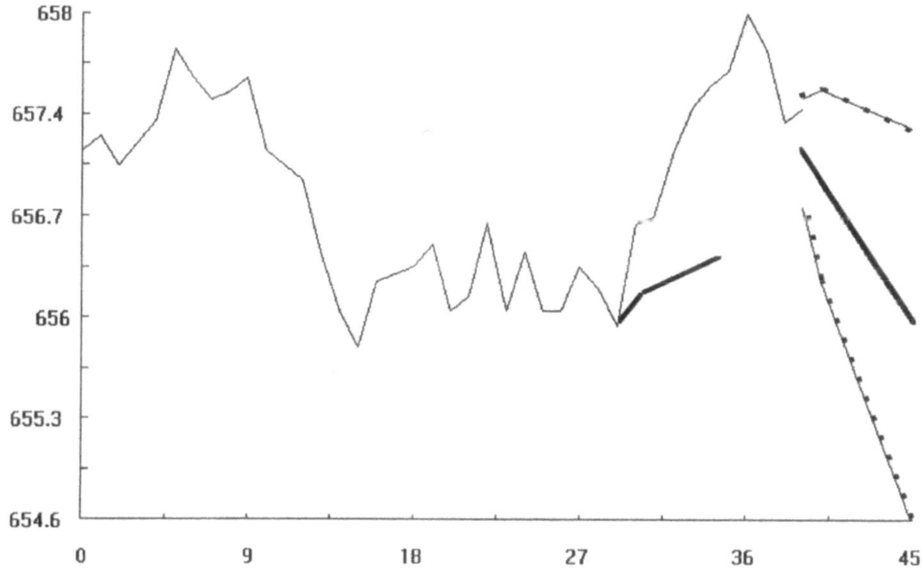

FIGURE 3.1.4.1 A Graphical Display Example (First Frame)
(Courtesy of Rapid Clip Neural Systems, Inc.)

observed since then. In this case, the figure shows that a downward trend was predicted at that time and a downward trend actually occurred after it was predicted.

During concurrent operation, the graphical prediction display is updated occasionally. Figure 3.1.4.2 shows an updated display at minute 49, 10 minutes after the Figure 3.1.4.1 display. Like its predecessor, Figure 3.1.4.2 includes a forecast telescope beginning at the current time point. Figure 3.1.4.2 also includes a thick line indicating its predicted values 10 minutes earlier, beginning at minute 39. These are the same as the current forecast values that were predicted in Figure 3.1.4.1, but at this point the previously forecast values have been actually observed. As the figure shows, the upward trend prediction from the earlier figure has been supported by index values observed since then. In this way graphical display proceeds continuously, providing the user an ongoing visual display of recent prediction accuracy.

Users can specify different graphical displays for different applications [1], as summarized in the Figure 3.1.4.3 specification list. As with other specification lists in this chapter, different options for a given specification are indented in the figure. For example, two options fall under the "Output Display Frequency" specification, and two options fall under the "Output Display Scale Information" specification. Also, each option may have several alternatives itself. For example, the "Time" option has four options under it. Some specifications are indica-

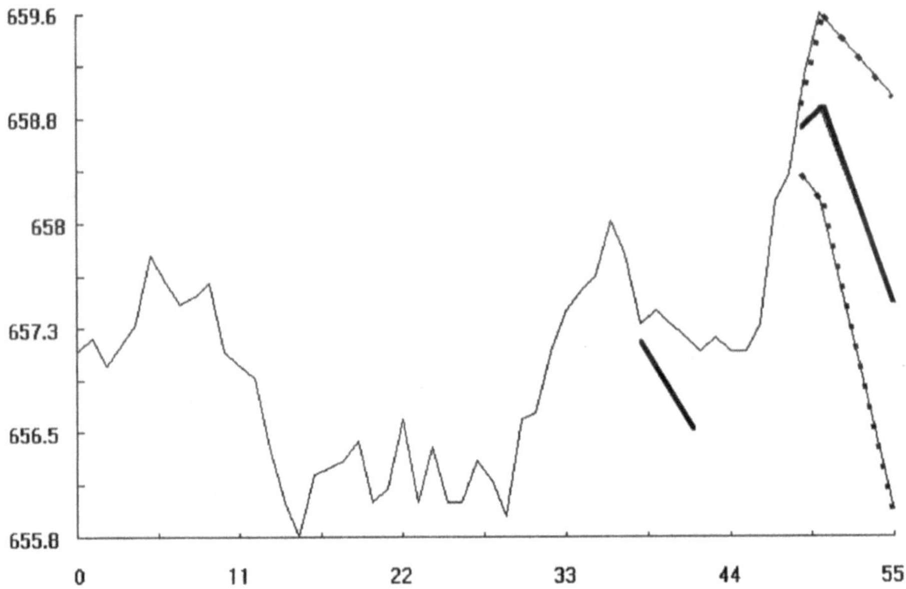

FIGURE 3.1.4.2 A Graphical Display Example (Second Frame)
(Courtesy of Rapid Clip Neural Systems, Inc.)

ted as "Optional," meaning that if an option is not selected then a default option will automatically be selected. For example, if one "Historical Prediction Plotting" option were not explicitly selected, then "Display Predicted Values at the Last Update Only" would be automatically selected.

Each of the Figure 3.1.4.3 options will be described next. Beginning with the "Output Display Frequency" option, a new plot will be displayed at every time point if the default is chosen. To update displays less often than every time point, the alternative option ("Skip Time Points Between Display Updating") is chosen and a "Time Point Skip Value" entry is supplied. For example, a skip value of 10 would produce Figure 3.1.4.1 and Figure 3.1.4.2 consecutively.

The "Output Display Scale Information" specifications give the user horizontal and vertical scaling alternatives. The "Value (Vertical) Scale" default specification produces sliding scales that adjust minimum and maximum values from trial to trial. This specification produces graphs that may jump from trial to trial in a distracting way in some applications. In that case, the user may override the default by specifying fixed minimum and maximum values.

The "Time (Horizontal) Scale" default specification produces sliding time scale displays. The user can modify the default specification to adjust the number of time points in each graph. The user can also choose telescoping scales like

- Output Display Frequency (Choose One — Optional)
  - Update at Every Time Point (The Default)
  - Skip Time Points between Display Updating
    - Time Point Skip Value: [Any Integer Greater Than One]
- Output Display Scale Information (Optional)
  - Value (Vertical) Scale (Choose One — Optional)
    - Use a Sliding Scale (The Default)
    - Use a Fixed Scale:
      - Minimum Value: [Any Number]
      - Maximum Value: [Any Number]
  - Time (Horizontal) Scale (Choose One — Optional)
    - Use a Sliding Scale (The Default)
      - Scale Size (Choose One — Optional)
        - Display 50 time points (The Default)
        - Display other than 50 time points: [Any Positive Integer]
    - Use a Sliding Scale Starting after Zero
      - Starting Value: [Any Integer]
      - Scale Size (Choose One — Optional)
        - Display 50 time points (The Default)
        - Display other than 50 time points: [Any Positive Integer]
    - Use a Telescoping Scale Starting At Zero
    - Use a Telescoping Scale Starting After Zero
      - Starting Value: [Any Integer]
  - Historical Prediction Plotting (Choose One — Optional)
    - Display Predicted Values at the Last Update Only (The Default)
    - Display Predicted Values at Other Time Points
      - Number of Historical Prediction Time Points: [Any Positive Integer]
      - Time Lag Between Time Points: [Any Positive Integer]

FIGURE 3.1.4.3 Graphical Specifications
(Courtesy of Rapid Clip Neural Systems, Inc.)

Figure 3.1.4.1 and Figure 3.1.4.2, which produce graphs that always begin at a fixed time point but include all time points from the fixed point to the current time point.

The default "Historical Prediction Plotting" option produces only one previous prediction set per graph — like the thick line beginning at time point 29 in Figure 3.1.4.1. The user may choose to have more than one prediction set, or adjust the time span between prediction time points, by choosing the indicated alternatives to the default option.

Figure 3.1.4.4. shows the *Rapid Learner*™ dialog box for graphical display [1]. Each section of the dialog box corresponds to one of the main specifications

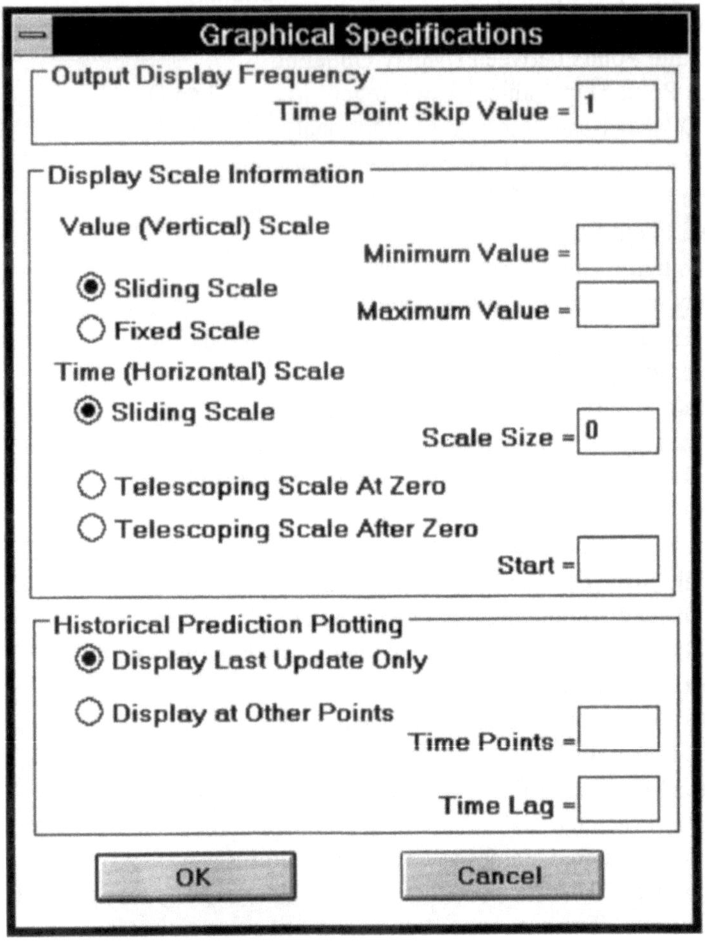

FIGURE 3.1.4.4 Graphical Specification Display
(Courtesy of Rapid Clip Neural Systems, Inc.)

in its corresponding Figure 3.1.4.3 list. Like other specification displays, this dialog box appears on the user's computer monitor during preliminary setup operations. When a new project is being prepared, the box will show all specifications set at their default values. When an existing project is being edited, the dialog box will show all specifications as they are currently set. The specifications in Figure 3.1.4.4 are set at their default values in keeping with Figure 3.1.4.3 as follows:

- "Time Point Skip Value" is set to 1.

- "Sliding Scale" is selected under "Value (Vertical) Scale."
- "Sliding Scale" at zero is selected under "Time (Horizontal) Scale."
- "Display Last Update Only" is selected under "Historical Prediction Plotting."

3.1.5 Operation Specification Overview

Along with graphical display specifications, the user may specify a variety of other *Rapid Learner*™ operations, which are summarized in Figure 3.1.5.1. "Data Input Specifications" include the number of input measurements and their types. "Learning Specifications" govern learning weight schedules, which in turn govern the relative learning impacts of measurements at different time points (see section 5.2 and chapter 6). Monitoring specifications govern the measurements to be monitored, their tolerance bands, and their deviance statistics (see section 7.1). Forecasting specifications govern the number of points to forecast and their tolerance bands (see section 7.2). Control specifications govern output predicted measurement values and related statistics that are provided to control external processes (see section 8.1). Refinement specifications govern the frequency and nature of feature function modifications (see section 8.2). Feature specifications govern the measurement functions that are used for prediction (see chapter 6).

Each specification in Figure 3.1.5.1 is summarized later in this chapter. As in the above graphical specification summary, a pair of figures is used to summarize each specification — one listing its options and one illustrating its *Rapid Learner*™ windows dialog box.

3.1.6 Data Input Specifications

Figure 3.1.6.1 shows input specifications including the number of input <u>records</u> (time points), the number of input measurements per record, input measurement types, the measurement source, <u>plausibility</u> status, and feature function types. Feature function types are described in a separate section (see section 3.6), but all

Data Input Specifications	[See Figure 3.1.6.1]
Learning Specifications	[See Figure 3.1.7.1]
Monitoring Specifications	[See Figure 3.2.2.1]
Forecasting Specifications	[See Figure 3.3.2.1]
Control Specifications	[See Figure 3.4.2.1]
Refinement Specifications	[See Figure 3.5.1.1]
Feature Specifications	[See Figure 3.6.2.1]

FIGURE 3.1.5.1 Operational Configuration Specifications
(Courtesy of Rapid Clip Neural Systems, Inc.)

Number of Input Records (Optional — Choose One)
  All Records (The Default)
  User-Specified Number of Input Records: [Any Positive Integer]
Record Skipping (Optional — Choose One)
  Do Not Skip Records (The Default)
  Skip Records
    Skip Size [Any Positive Integer]
Number of Input Measurements per Record: [Any Positive Integer]
Input Measurement Types (Optional — Choose One)
  Arithmetic Only (The Default)
  Binary Only
  Categorical Only: [Category Count List]
  Mixed Types: [Measurement Type List]
Input Measurement Source (Optional)
  User-Supplied File (The Default)
  System Generated: [Measurement Generation Specifications]
  User-Supplied Function
Measurement Plausibility Status (Optional — Choose One)
  No Measurements Missing (The Default)
  Plausibility Values Supplied
    Accurate Prediction (The Default)
    Precise Prediction
Measurement Feature Specifications (Optional) [see Figure 3.6.2.1]

FIGURE 3.1.6.1 Data Input Specifications
(Courtesy of Rapid Clip Neural Systems, Inc.)

other input specifications are explained below.

During concurrent operation, the *Rapid Learner*™ simulator reads measurements from a file containing each measurement value in fixed format and separated by one or more blanks from adjacent measurement values. Measurement values are grouped into records with one record per time point. Unless the user specifies the "Number Of Input Records" with a positive integer, the simulator continues concurrent operation until it finds no more measurement values in the input file.

In some applications, more records may be available than are necessary for concurrent learning and performance. The "Record Skipping" specification, which is designed for such applications, allows the user to skip one or more measurements between each measurement record.

The user must specify the "Number of Input Measurements per Record." The user may also optionally specify "Input Measurement Types" as either arithmetic, binary, or categorical (see sections 6.1 and 6.5). If "Mixed Types" is selected, the user supplies a list of category types, with one type listed for each measurement [1]. If categorical measurement types are listed, the user specifies the number of categories in each type (see section 6.5).

The "Input Measurement Source" option specifies the measurement records that are processed by *Rapid Learner*™ software. For monitoring and forecasting simulation, measurement records may be created prior to concurrent operation. They may be either in a "User Supplied File" attached to the Client Output File (see section 3.1.3), "System Generated" by *Rapid Learner*™ data simulation functions [1], or generated by a "User-Supplied Function." If measurements are), "System Generated," a variety of other options, not shown in Figure 3.1.6.1, must be specified.

Input records that are generated by a "User-Supplied Function" are useful when simulating rapid learning during control (see section 8.1). For control simulation, input measurements may be more difficult to represent than in monitoring and forecasting simulation, because concurrent control signals may affect future input measurements (see sections 3.6 and 8.1). The best way to assess rapid learning performance in such settings is to install a real-time rapid learning system directly into a process. The best simulation alternative is to generate simulated future measurement records that depend on *Rapid Learner*™ output control signals from trial to trial. That alternative can be selected by choosing the "User-Supplied Function" measurement source option, which requires that the user supply an executable subroutine to the Client Output File (see section 3.1.3). *Rapid Learner*™ simulation software is programmed to send output control data to that subroutine at the end of each trial and receive an input measurement record from it at the beginning of the next trial [1].

The "Measurement Plausibility Status" specification allows users to treat missing measurement values in different ways. Each input record contains a fixed field for all measurement values, whether they are missing or not. All values are assumed to be not missing if the "No Measurements Missing" option is selected. If the "Plausibility Values Supplied" option is selected, however, measurement values may be missing from trial to trial. In that case, the user supplies a plausibility record for each measurement record to indicate which measurements are missing (see section 4.1.1). If more than one measurement is missing in a given record, predicting that measurement precisely requires making a time-consuming adjustment to the connection weight matrix (see section 8.2.2). The user has the alternative of avoiding this adjustment by selecting the "Fast Prediction" option, if speed is more important than accuracy.

Figure 3.1.6.2 shows the *Rapid Learner*™ dialog box for all input specifications, except feature function options, which have their own dedicated dialog box (see section 3.6). The specifications are set at their default values in keeping with Figure 3.1.6.1, as follows:

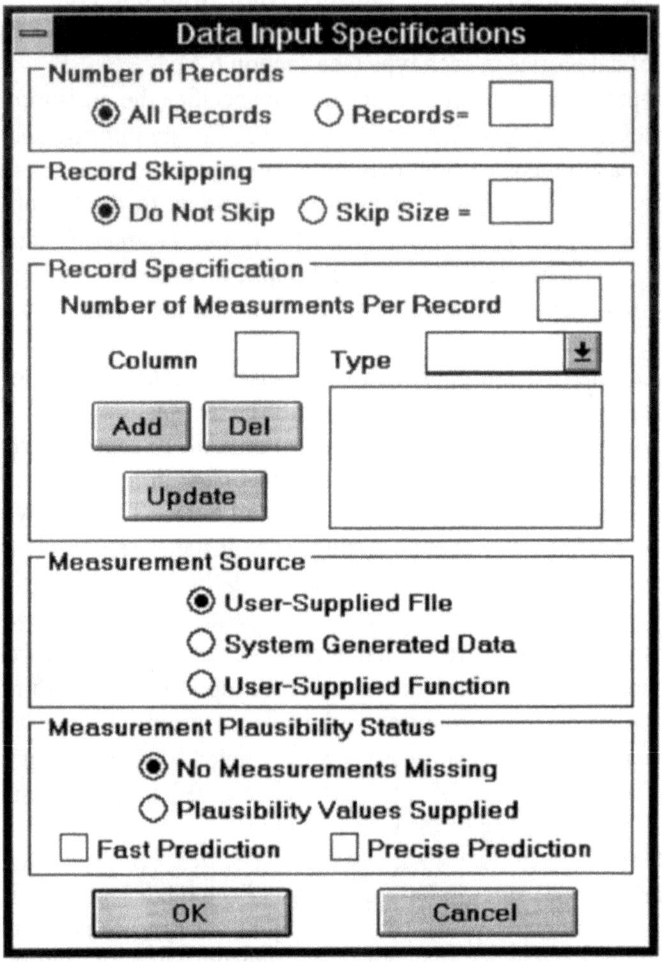

FIGURE 3.1.6.2 Data Input Specification Display
(Courtesy of Rapid Clip Neural Systems, Inc.)

- The "Number of Records" entry is set to "All Records."
- "Do Not Skip" is selected "Record Skipping."
- The "Number of Measurements Per Record" is blank — this entry must be made by the user.
- A measurement "Type" list is provided that assigns type A (for arithmetic), B (for binary), or C (for categorical) to each measurement.
- The "User-Supplied File" is selected under the "Measurement Source" option.

- "No Measurements Missing" is selected under the "Measurement Plausibility" option.

3.1.7 Learning Specifications

Learning specifications govern the relative impacts of measurement records on prediction learning (see section 5.2.3). One learning specification option is based on grouping measurement records into blocks. Each record in each block has the same learning impact as other measurements in the same block. Each block has an overall impact that is a specified proportion of all learning that has occurred up to and including the block. For example, suppose that block length is set to 50 and block impact is set to 0.5. Then at the end of 600 trials, the block that began at time point 551 and ended at time point 600 will have produced 50% of the overall learning impact up to time point 600, and each of the 50 trials within that block will have produced 1% of the overall impact. The previous 11 blocks would have similar learning impacts relative to their ending time point. The overall effect is block learning impact levels that decrease exponentially from the most recent time block onward (see section 5.2.3).

As an alternative to blocked learning, the user may specify equal learning impact for each record. In that case, the user supplies an equal impact learning factor. The learning factor is the amount of learning impact from each measurement record, relative to impact from initialized learned parameters (see section 5.2.1). For example, suppose that a factor of 2 is chosen. Then after one trial the first measurement record will have two impact units, prior learning will have one impact unit, and the first record will have 2/3 of the overall impact (2/3 = 2/(1+2); after two trials, the first and the second measurement records will each have two impact units and 2/5 of the overall impact (2/5 = 2/(1+2+2); after three trials the first, second, and third measurements will each have 2/7 of the overall impact (2/7 = 2/(1+2+2+2); and so on. In this way, the sum of initial and all subsequent impact values will be 1 at every time point, and each observation will have twice the initial impact.

The user may also tailor learning impact to fit specific application needs by supplying learning weights from a file (see section 5.2.3). In addition, adaptive monitoring, control, and refinement applications may require variable learning impact values that depend on recent monitoring results (see chapter 7, 8 and 9). In that case, the user may supply a learning weight control subroutine, which supplies *Rapid Learner*™ software a learning weight value at the beginning of every concurrent trial.

Figure 3.1.7.1 summarizes user-supplied learning specification options. "Blocked Learning" is the default, with a "Most Recent Block Impact Value" of 0.5 and a "Block Length" value of 100. If "Equal Impact Learning" is chosen, then a "Learning Factor" value of 1.0 is the default. If "User-Supplied Learning

Learning Specifications (Optional — Choose One)
Blocked Learning (The Default)
Most Recent Block Impact Value (Optional — Choose One)
0.5 (The Default)
User-Specified Impact Value: [Any Positive Number Less Than 1.0]
Block Length (Optional — Choose One)
100 (The Default)
User-Specified Block Length: [Any Positive Integer]
Equal Impact Learning
Learning Factor Value (Optional — Choose One)
0.5 (The Default)
User-Supplied: [Any Positive Number]
User-Supplied Learning Weights (Choose One)
Learning Weights Supplied from a File (The Default)
Learning Weights Generated by a User-Supplied Function

FIGURE 3.1.7.1 Learning Specifications
(Courtesy of Rapid Clip Neural Systems, Inc.)

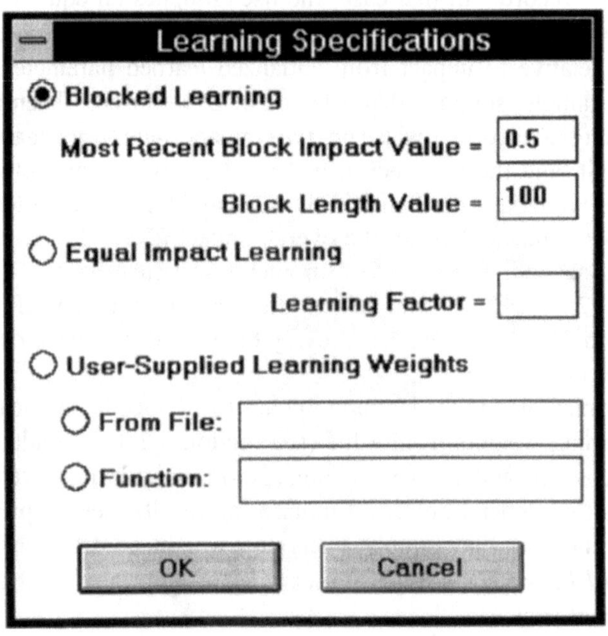

FIGURE 3.1.7.2 Learning Specification Display
(Courtesy of Rapid Clip Neural Systems, Inc.)

Weights" is chosen, then the user controls learning impact by either supplying a learning weight file (the default) or an executable subroutine to the Client Output File (see section 3.1.3).

Figure 3.1.7.2 shows the *Rapid Learner*™ dialog box for learning specifications. The specifications are set at their default values in keeping with Figure 3.1.7.1, as follows:

- The "Blocked Learning" entry is selected.
- The "Most Recent Block Impact" value is set to 0.5.
- The "Block Length" value is set to 100.

3.2 MONITORING OPERATION

This section gives *Rapid Learner*™ software monitoring operation details, in terms of a strain gauge monitoring example (see section 3.2.1). Several monitoring specifications are listed here and explained briefly as an operational introduction, which is extended to detailed description later in the book (see section 7.1). Monitoring statistic options are introduced in the same way (see section 3.2.2, 7.1.2, and 7.1.3).

3.2.1 A Strain Gauge Monitoring Example

A degree of user control over monitoring specifications is useful in most applications. For example, when measurements from several hundred strain gauges are available (see section 1.1.2), only a proportion of them may require concurrent monitoring. *Rapid Learner*™ software allows a subset of all measurements to be selected for concurrent monitoring. Only graphs for the selected subset are provided, with provision for selecting other graphs if necessary.

A second aspect of monitoring where alternatives are useful is the choice of tolerance bands. Tolerance bands are based on learned mean squared differences between observed and predicted measurement values, as well as on standard results from statistical theory. *Rapid Learner*™ monitoring software sets default tolerance band values to ±2 standard error units, based on these mean squared differences. (see section 7.2.1). If standard statistical assumptions are met, this setting will produce false alarms at a frequency of about 5%, which is acceptably small in many applications. In other applications, however, monitoring alarms based on narrower tolerance bands may be needed for higher monitoring sensitivity. On the other hand, high sensitivity may also result in high false alarm rates, which may pose problems in still other applications. In such cases, setting wider tolerance bands may be more appropriate.

During normal monitoring operation, *Rapid Learner*™ software routinely provides a display of observed values, predicted values, and tolerance bands for measurements that the user selects (see Figure 3.2.2.2 below). Special-purpose displays may be provided with *Rapid Learner*™ software to suit specific application needs as well (e.g. see Figure 1.1.2.3). In either case, audible alarms and visual displays may be useful for indicating monitored values that have recently exceeded their tolerance bands. *Rapid Learner*™ software gives users the option of specifying such alarms, in the form of an audible alarm and a list of measurements that have shown up as deviant at the current time point for the first time.

A third monitoring aspect where alternatives are useful is the treatment of deviant measurements. When deviant measurements are discovered during the monitoring process, using them as predictors for monitoring other measurements may or may not be useful, depending on the application. In some applications, deviant measurements may be sufficiently faulty to produce highly inaccurate predictions of other measurements. In that case, excluding deviant measurements entirely from future predictions of other variables may be useful. However, since excluding such deviant measurements requires computer time, including deviant measurements may be best in other applications, especially when the deviant values have minor prediction impact on other measurements and they tend to be only spurious. A fast, but potentially inaccurate, compromise for dealing with deviant measurements is to set viability values for deviant measurements to zero. Setting deviant viability values to zero has the overall effect of deleting terms from the prediction equation that involve the deviant measurements, without adjusting other weights properly (see section 5.2).

Along with graphical displays, *Rapid Learner*™ software provides a log of deviant monitoring events. Each log entry includes the deviance time point, the deviant measurement value, its predicted value, and several prediction statistics. These statistics may be useful in identifying prediction problems that may have produced the deviance between observed and predicted values. The statistics are provided in the form of a table, with each row of the table corresponding to an independent feature that was used to predict the deviant measurement value. Statistics that are provided in each row of the table include the independent variable value, its prediction coefficient, its raw contribution to predicting the deviant dependent variable value, its standardized prediction coefficient, and its standardized contribution to predicting the deviant dependent variable value (see section 7.1.2).

3.2.2 Monitoring Options

Figure 3.2.2.1 summarizes monitoring specifications, including a list of "Measurements to be Monitored," "Monitoring Tolerance Band" options, "Devi ant Measurement Exclusion" options, and a "Monitoring Statistics" option (see

Measurements to be Monitored (Optional — Choose One)
All (The Default)
Monitoring Variable List: [User-Specified List]
Monitoring Tolerance Band (Optional — Choose One)
Factor Value
2.0 (The Default)
User-Specified Tolerance Band Factor: [Any Positive Number]
Monitoring Alarm (Optional — Choose One)
Display an Alarm if Tolerance Band is Exceeded (The Default)
Do Not Display an Alarm
Deviant Measurement Exclusion (Optional — Choose One)
Set Deviant Viability Values to Zero (The Default)
Exclude Deviant Measurements Permanently
Include Deviant Measurements
Monitoring Statistics (Optional — Choose One)
Exclude (The Default)
Include

FIGURE 3.2.2.1 Monitoring Specifications
(Courtesy of Rapid Clip Neural Systems, Inc.)

section 7.1.2). All measurements are included for monitoring by default, but any subset of them may be included instead. The default tolerance band factor value of 2.0 may be replaced by any positive number, with smaller numbers imposing more stringent monitoring criteria and larger numbers imposing less stringent criteria. The monitoring alarm will be enabled by default, but it may be disabled. Deviant measurement values will be assigned viability values of zero by default, but they may instead be either excluded permanently or ignored and included. Permanent exclusion involves a connection weight adjusting process that may interrupt concurrent operation for a brief period (see section 8.2.2).

Figure 3.2.2.2 shows the *Rapid Learner™* dialog box for monitoring specifications. The specifications are set to their default values as follows:

- "All" is specified in the "Measurements Monitored" box.
- A "Tolerance Band" "Factor Value" of 2.0 is selected.
- The "Alarm if Deviant" box is checked.
- "Set Viabilities to Zero" is selected in the "Deviant Measurements" box.
- "Exclude" is selected in the "Monitoring Statistics" box

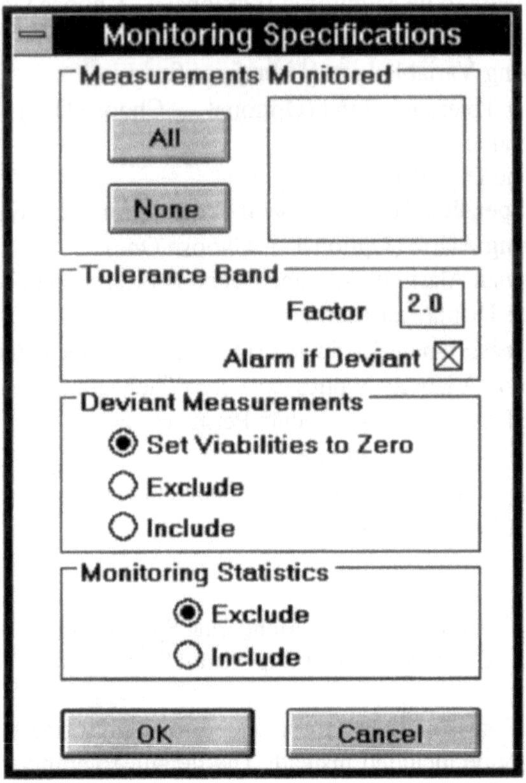

FIGURE 3.2.2.2 Monitoring Specification Display
(Courtesy of Rapid Clip Neural Systems, Inc.)

3.3 FORECASTING OPERATION

This section gives *Rapid Learner*™ software forecasting operation details, in terms of a commodities forecasting example (see section 3.3.1). A variety of forecasting specifications are listed here and explained briefly as an operational introduction, which is extended to detailed description later in the book (see section 7.2). Other forecasting options provided by the software are introduced in the same way (see section 3.3.2).

3.3.1 A Commodities Forecasting Example

Several options are useful for specifying a variety of forecasting needs. These include specifying which measurements to forecast, how many future time points to forecast, how many time points to skip between each time point, and how wide forecasting tolerance bands should be. Choosing forecast options requires careful thought when forecasting speed is critical, because concurrent processing computer time increases with the number of measurements to forecast and with the number of forecast time points per measurement. In some applications, such as forecasting the Standard & Poors Index, forecasting all predictor variables for the index may not be necessary. For other applications, such as making concurrent trading decisions on each of the component commodities for the index, forecasting each measurement may be necessary. In these applications, choosing only a few forecast time points per measurement and skipping several time points between each pair of forecast time points may be best.

Having optional control over forecasting tolerance band widths may be useful as well. With commodities forecasting, for example, being able to identify future trends with a relatively low confidence level, say 75%, may be sufficient for producing profitable trades in the long run. However with other applications such as predicting critical health care needs, a higher confidence level than 95% may be necessary. Several *Rapid Learner*™ tolerance band options for meeting these and other needs are described below.

3.3.2 Forecasting Options

Forecasting Specifications (Optional — Choose One)
- Do Not Perform Forecasting (The Default)
- Perform Forecasting
  - Number of Forecasting Time Points (Optional — Choose One)
    - One (The Default)
    - More Than One
      - Forecasting Time Point Count: [Any Integer Greater Than One]
      - Time Span Between Forecasting Points (Optional — Choose One)
        - One (The Default)
        - More Than One: [Any Integer Greater Than One]
  - Measurements to be Forecast (Optional — Choose One)
    - All (The Default)
    - Forecasting Variable List: [User-Specified List]
  - Forecasting Telescope Tolerance Band Factor (Optional — Choose One)
    - 2.0 (The Default)
    - User-Specified Tolerance Band Factor: [Any Positive Number]

FIGURE 3.3.2.1 Forecasting Specifications
(Courtesy of Rapid Clip Neural Systems, Inc.)

Figure 3.3.2.1 summarizes *Rapid Learner*™ software forecasting specifications. The default option, "Do Not Perform Forecasting," is appropriate for pure monitoring applications or simple control applications based on current deviance values only (see section 3.4). Forecasting options include "Number of Forecasting Time Points," "Measurements to be Forecast," and "Forecasting Telescope Tolerance Band Factor." If the "Number of Forecasting Time Points" is "More Than One," then any positive integer may be selected as the "Time Span Between Forecasting Points." The default "Forecasting Telescope Tolerance Band Factor" of 2.0, which may be interpreted as producing 95% confidence telescopes (see section 7.2), may be replaced with lower factor values with corresponding lower confidence levels or higher factor values with corresponding higher confidence levels.

Figure 3.3.2.2 shows the *Rapid Learner*™ dialog box for initial monitoring specifications. The specifications are set at their default values in keeping with Figure 3.3.2.1 as follows:

- Under "Measurements to Forecast," "All" is selected.

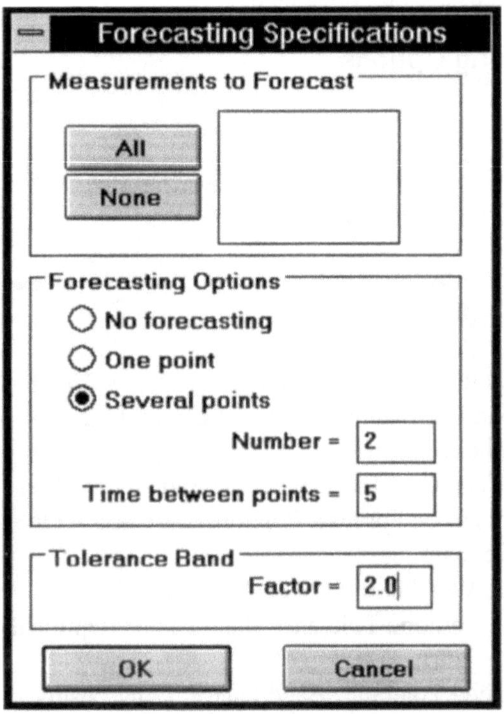

FIGURE 3.3.2.2 Forecasting Specification Display
(Courtesy of Rapid Clip Neural Systems, Inc.)

- The "No Forecasting" entry is selected — if the "Several points " box is selected, a default value of 2 will appear in the "Number = " box and a default value of 1 will appear in the "Time between points = " box.
- A "Tolerance Band" factor of 2.0 is selected.

3.4 CONTROL OPERATIONS

This section gives *Rapid Learner*™ software control operation details, in terms of a missile tracking example (see section 3.4.1). A variety of control specifications are listed here and explained briefly as an operational introduction, which is extended to detailed description later in the book (see section 8.1). Other forecasting options provided by the software are introduced in the same way (see section 3.4.2).

3.4.1 A Missile Tracking Control Example

Referring to the missile tracking example that was presented earlier (see section 1.3.1), controlling a tracking camera or a homing device requires establishing a missile position prediction function. *Rapid Learner*™ software establishes prediction functions for control through user-supplied feature specifications (see section 3.6). For example, missile tracking control may be described in terms of predicting three missile position coordinates in the next time frame as a function of related current and recent measurements. Such measurements might include missile coordinate and attitude measurements from separate sensing devices and missile thrust data from other sensors. Current values as well as a recent history of such measurements may be combined in a variety of ways, in order to reflect a variety of linear and non-linear relationships (see chapter 6). For purposes of this example, suppose that measurements are combined to form 50 prediction features. Suppose further that 53 total variables are available for control at each time point: the 50 prediction features and the three missile coordinate values (latitude, longitude, and height).

Besides establishing input-output relationships for control, *Rapid Learner*™ software can supply a variety of output signals for control. Output control signals may include any predicted values and their standard errors (see sections 3.2 and 5.2), for any variables in the system. In the simplest case, the output control signal for any given variable will be the forecast value for that variable one time point into the future. More sophisticated control devices may utilize forecast values several points into the future along with their standard errors, in order to ensure smooth control operation (see chapter 8.1). For purposes of this example, suppose that at each time point an external camera tracking device requires forecast values of the three position coordinates (variables 51-53) at the next time point.

Control signals have no effect on future measurement values in relatively simple settings. For example, when control signals are used to keep a camera trained on a missile, the signals have no effect on future missile position coordinates and missile thrust values. In many other settings, however, the control signals are highly related to future measurement values. For example, when control signals are used to keep a homing device on a missile then the missile may take evasive action accordingly, substantially affecting future input values. Simulations are relatively simple to perform in the former, so-called open-loop case (see section 8.1.2) because all input variable measurements may be generated prior to the simulation. In the latter, closed-loop case, however, each new measurement record is produced as a function of current and recent control signals. The *Rapid Learner*™ Simulator allows users to generate future measurements concurrently by supplying subroutines for measurement generation from trial to trial (see section 3.1.4). Subroutine inputs include output control signals from the current trial, and subroutine outputs include simulated measurement inputs for the next trial [1].

3.4.2 Control Options

Figure 3.4.2.1 lists *Rapid Learner*™ control specifications, including whether or not to "Supply Output Control Signals," a "Control Variable List" for which control signals will be supplied and an "Output Control Data" specification. If "Near-Term Forecast Value Only" is selected, only predicted values for the selected control variables at the next time point are provided. If "Near-Term Forecast Value and its Tolerance Band Only" is selected, the predicted value for each control variable is followed by its current standard error value. Otherwise, "All Imported and Forecast Values and their Tolerance Bands" are provided.

Control Specifications (Optional — Choose One)
- Do Not Supply Control Signals (The Default)
- Supply Control Signals
  - Control Variable List (Optional — Choose One)
    - The Last Variable Only (the default)
    - All
    - Some [Variable List]
  - Output Control Data (Optional — Choose One)
    - The Near-Term Forecast Value Only (The Default)]
    - The Near-Term Forecast Value and its Tolerance Band Only (The Default)]
    - All Imputed and Forecast Values and their Tolerance Bands

FIGURE 3.4.2.1 Control Specifications
(Courtesy of Rapid Clip Neural Systems, Inc.)

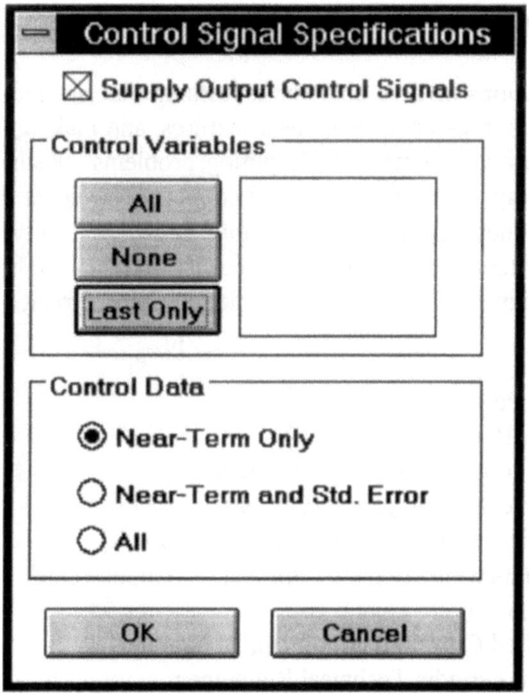

FIGURE 3.4.2.2 Control Specification Display
(Courtesy of Rapid Clip Neural Systems, Inc.)

Figure 3.4.2.2 shows the *Rapid Learner*™ dialog box for initial monitoring specifications. The specifications are set at their default values in keeping with Figure 3.4.2.1 as follows:

- "Supply Output Control Signals" is not selected.
- Only the last variable in the list is included.
- "Near-term only" is selected under the "Control Data" option.

3.5 REFINEMENT OPERATIONS

This section gives *Rapid Learner*™ software refinement operation details, in terms of a visual pattern recognition example (see section 3.5.1). A variety of control specifications are listed here and explained briefly as an operational introduction, which is extended to a detailed description later in the book (see section 8.2). Other forecasting options provided by the software are introduced in the same way (see section 3.5.2).

3.5.1 Refinement Options

Refinement operations involve combining features that are redundant, removing features that are unnecessary, adding new features, and making technical parameter corrections to avoid numerical computing problems. Refinement operations require periodic inspection of inter-feature connection weights and other parameters that are continuously updated by *Rapid Learner*™ software. Refinement options specify criteria for parameter inspection frequency, technical adjustments, redundancy adjustments, unnecessary prediction feature removal, and new feature inclusion.

Refinement Time Period (Optional)
  Every 100 Time Points (The Default)
  User-Specified Period: (Any Positive Integer)
Technical Refinement Specification (Optional — Choose One)
  Perform Periodic Technical Refinement (The Default)
    Technical Refinement Criterion (Optional — Choose One)
      32 (The Default)
      User-Specified Criterion: [Any Positive Number]
  Do Not Perform Periodic Technical Refinement
Redundancy Refinement Specification (Optional — Choose One)
  Do Not Perform Redundancy Refinement (The Default)
  Perform Redundancy Refinement
    Redundancy Criterion (Optional — Choose One)
      Distance Value of 0.01 (The Default)
      User-Specified Distance Value: [Any Positive Number Less Than 1]
Predictability Refinement Specification (Optional — Choose One)
  Do Not Perform Predictability Refinement (The Default)
  Perform Predictability Refinement (Multiple Use of This Option is Permitted)
    Dependent Variable: [Any Positive Integer up to the Number of Features]
    Predictability Criterion (Optional — Choose One)
      Predictability Factor of 0.01 (The Default)
      User-Specified Predictability Factor: [Any Positive Number Less Than 1]
Feature Replacement Specification (Optional — Choose One)
  Do Not Insert New Features (The Default)
  Insert New Features (Optional — Choose One)
    Insert User-Supplied Features (The Default)
    Insert Recent Features (The Default)
    Insert Power Features to Replace Excluded Features

FIGURE 3.5.1.1 Refinement Specifications
(Courtesy of Rapid Clip Neural Systems, Inc.)

Figure 3.5.1.1 summarizes refinement specification options. The "Refinement Time Period" specifies how often refinement operations occur. This parameter may be critical when performing redundancy refinement, where required computer time increases exponentially with feature set size (see section 5.2). As a result, if single-processor systems are used (see section 3.1), concurrent operations may be interrupted for long time periods while refinement operations proceed. Interrupt times are shorter for other forms of refinement. In many applications, data records arrive seldom enough so that refinement could proceed between every trial without interruption. However, making refinement adjustments more often than every 100 trials, the default "refinement period" is seldom necessary.

The "Technical Refinement Specification" in Figure 3.5.1.1 sets a criterion for avoiding numerical overflow problems. Such problems occur when measurements have very low variances and very high correlations (see section 8.2). For example, routine deviance value computing requires division by learned standard errors (see section 7.1), which approach zero over time if measurements do not vary. *Rapid Learner*™ software adjusts standard errors in this case by adding to them small positive values that avoid dividing by zero. Technical refinement involves checking the size of certain variance and connection weight value exponents (see sections 8.2.4 and 8.2.5). If exponent values exceed 32 by default, *Rapid Learner*™ software makes appropriate adjustments.

The "Redundancy Refinement Specification" in Figure 3.5.1.1 sets a criterion for replacing closely related features by their average. "Clustering" features in this way reduces computing time and space requirements, and it avoids numerical problems (see section 8.2.1). During redundancy checking, a redundancy distance measure is computed for each pair of features. The distance measures how dissimilar the pair members are in terms of predicting all other features (see section 8.2.1). A value of zero indicates that they are completely equivalent, while a high value indicates that they are completely independent.

The "Predictability Refinement Specification" governs the removal of features that are unnecessary for prediction. In many applications, one or more features have fixed independent variable roles and the others have fixed dependent variable roles. For example, missile tracking places next-position coordinates in an independent variable role and all other features in a dependent variable role (see section 8.1). In such applications, identifying independent variables that are unnecessary for improving prediction accuracy may be useful. Removing independent variables increases processing speed, decreases processing memory requirements, and leaves room for adding new features that may improve predictability. The "Predictability Factor" removal criterion is the tolerance band reduction proportion that is achieved by including a given independent variable (see section 8.2.2). A value of 0 for the criterion indicates that the predictor in question does not reduce the tolerance band at all, over and above other independent variables.

The "Feature Replacement Specification" allows new features to be inserted in place of features that have been combined or removed. Replacement features may be recent features, products among features, or new user-supplied feature specifications. Recent and product replacement features are inserted automatically by *Rapid Learner*™ software in ways that are described elsewhere [1], while user-supplied features are inserted according to modified feature function specifications (see section 3.6).

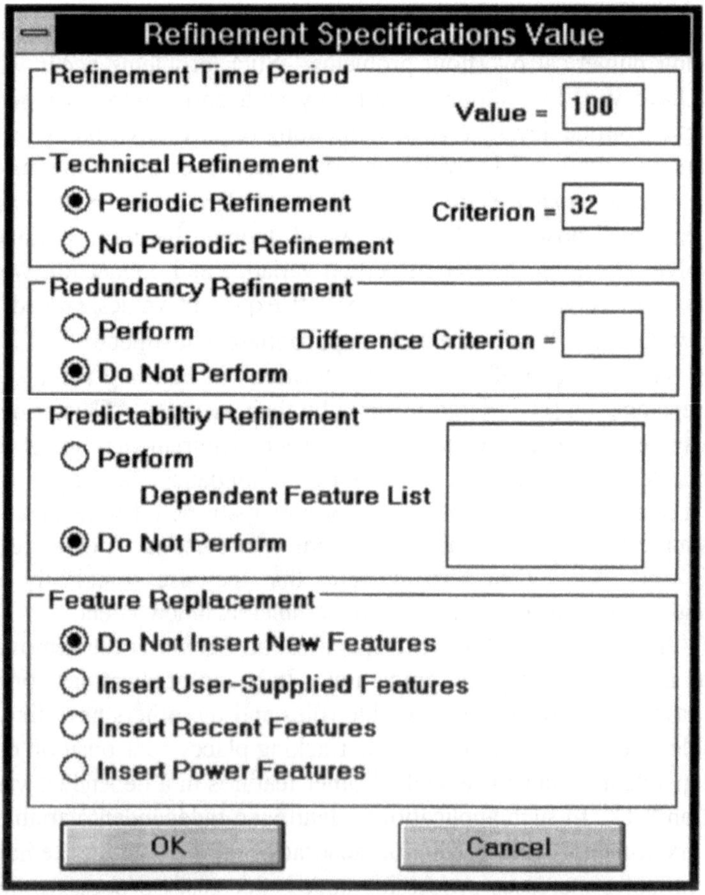

FIGURE 3.5.1.2 Refinement Specification Display
(Courtesy of Rapid Clip Neural Systems, Inc.)

Figure 3.5.1.2 shows the *Rapid Learner*™ dialog box for initial refinement specifications. The specifications are set at their default values in keeping with Figure 3.4.2.1, as follows:

- The "Refinement Time Period" value is set to 100.
- "Periodic refinement" is selected under "Technical Refinement", along with the default "Criterion" value of 32.
- "Do Not Perform" is selected under "Redundancy Refinement" — if "Perform" were selected, the default "Criterion" value of 0.01 would apply.
- "Do Not Perform" is selected under "Predictability Refinement" — if "Perform" were selected, the default "Criterion" value of 0.01 would apply to all features as independent variables, when used to predict the dependent variables in the "Dependent feature list".
- "Do Not Insert New Features" is selected under "Feature Replacement."

3.5.2 Feature Refinement Statistics

As a part of refinement operations, *Rapid Learner*™ software routinely produces a variety of feature function statistics, many of which can be informative to users. These include global statistics that describe individual and overall feature interrelation strengths. They also include prediction statistics that indicate how well specific input features predict specific output features. Users may examine any subset of these feature function statistics at every refinement time point (see section 3.5.1). This section introduces these statistics and shows how users may obtain them. Detailed descriptions and examples of individual statistics are provided later in the book (see section 8.2.5).

Figure 3.5.2.1 lists global refinement statistics that are provided by *Rapid Learner*™ software. All such statistics are values that are learned up to the current time point. Feature "Means" are weighted feature averages and "Standard Deviations" are measures of variation between feature values and their learned means. "Standard Errors" are measures of variation between observed and predicted feature values. Each of the feature "Tolerance Band Ratios" is a tolerance

Means
Standard Deviations
Tolerance Band Ratios
Pair-Wise Tolerance Band Ratios
Pair-Wise Partial Tolerance Band Ratios

FIGURE 3.5.2.1 Global Refinement Statistics
(Courtesy of Rapid Clip Neural Systems, Inc.)

band reduction factor, obtained by dividing the feature standard error by the feature standard deviation.

"Pair-Wise Tolerance Band Ratios" are also provided. The entry in a given row and a given column is the tolerance band reduction that may be obtained by predicting the row feature from the column feature and *vice versa*. Pair-wise tolerance band ratios are derived from standard descriptive statistics called product-moment correlation coefficients (see section 5.1).

"Pair-Wise Partial Tolerance Band Ratios" are provided in a table (see sections 8.2.2 and 8.2.5). The entry in a given row and a given column is the tolerance band reduction that may be obtained by predicting the row feature from the column feature and *vice versa,* as with pair-wise tolerance band ratios. In this case, the ratios reflect tolerance band reductions over and above tolerance band values that could be obtained using all other features as predictors. Partial tolerance band ratios are derived from standard descriptive statistics called partial correlation coefficients.

Turning next to prediction refinement statistics, the user may specify prediction refinement statistics for each feature variable in the feature set. For each selected feature variable, *Rapid Learner*™ software provides a table of statistics for predicting the selected feature as an independent variable, depending on all other features as independent variables. Each row of the table provides prediction refinement statistics for a distinct independent variable.

Figure 3.5.2.2 lists prediction refinement statistics that are provided for each independent variable. As with global statistics, prediction statistics are values that are learned up to the current time point. Each independent variable feature is used to predict a given dependent variable feature as part of a weighted sum. The "Raw Prediction Weights" are the coefficients for each independent variable. The "Standardized Prediction Weights" are values that coefficients for each independent variable would have if all features had a standard deviation value of 1.0. The "Tolerance Band Reduction Factors" are identical to the "Pair-wise Tolerance Band Ratios" in Figure 3.5.2.1, but are rearranged for easier interpretation. Likewise, the "Partial Tolerance Band Reduction Factors" are the same as "Pair-Wise Partial Tolerance Band Ratios" in Figure 3.5.2.1.

Figure 3.5.2.3 shows the *Rapid Learner*™ dialog box for initial refinement statistics. The specifications are set at their default values, as follows:

Raw Prediction Weights
Standardized Prediction Weights
Tolerance Band Reduction Factors
Partial Tolerance Band Reduction Factors

FIGURE 3.5.2.2 Prediction Refinement Statistics
(Courtesy of Rapid Clip Neural Systems, Inc.)

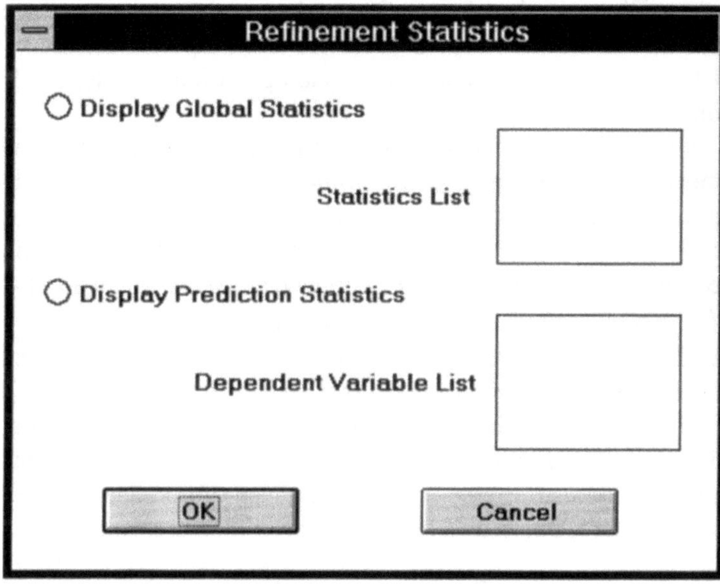

FIGURE 3.5.2.3 Refinement Statistics Specification Display
(Courtesy of Rapid Clip Neural Systems, Inc.)

- The "Display Global Statistics" entry is not checked — if it were checked, highlighted global refinement statistics in the "Statistics List" would be provided.
- The "Display Prediction Statistics" entry is not checked — if it were checked, a table of prediction refinement statistics would be provided for each feature included in the "Dependent Variable List."

3.6 FEATURE FUNCTION PROCESSING OPERATIONS

This section gives *Rapid Learner*™ software feature function specification operation details. The section begins with an operational overview of feature functions (see section 3.6.1), which are explained in detail later in the book (see chapter 6). Feature function processing options are introduced in the same way (see section 3.6.2).

3.6.1 Feature Function Overview

Rapid Learner™ software converts measurements to functions of measurements called features (see section 3.1, chapter 6). These feature functions, which may be linear or non-linear, determine the prediction functions that are used for all

Rapid Learner™ monitoring, forecasting, and control operations. Once features are computed, each feature value is predicted from all other feature values (see section 3.1, chapter 5). As a result, prediction precision may depend heavily on feature function selection.

The simplest way to specify features is to equate input features with input measurements. Equating measurements with features is accomplished by one-to-one-functions of purely arithmetic measurements (see section 3.1.6), but with binary or categorical measurements conversion to equivalent arithmetic features is required routinely (see section 6.5). These conversions are made automatically by *Rapid Learner*™ software without the need for user specifications.

Rapid Learner™ software provides a variety of other feature function computing alternatives for less simple cases (see section 3.6.2, chapter 6). The most general options are in the form of user-supplied subroutines, which can be programmed to produce any feature functions from input measurements. More specific options are also offered, including the use of non-linear power features and historical features based on a recent history of measurements.

3.6.2 Feature Function Processing Options

Figure 3.6.2.1 summarizes *Rapid Learner*™ feature function specification options. If no "Input Measurement Exclusions" are specified, then all input measurements are automatically used to compute feature functions. If the "Exclusion Specification" option is chosen, then measurements appearing in the "Measure-

Measurement Feature Specifications (Optional)
- Input Measurement Exclusions (Optional)
  - None (The Default)
  - Exclusion Specification: [Measurement Exclusion List]
- User-Supplied Subroutines (Optional)
  - None (The Default)
  - Number of User-Supplied Features: [Any Positive Integer]
- Mean Features (Optional)
  - Not Supplied (The Default)
  - Mean Feature Specification: [Cluster Indicator List]
- Historical Features (Optional)
  - Number of Historical Steps: [Any Positive integer]
  - Historical Step Size: [Any Positive Integer]
  - Power Degree: [Any Non-Negative Integer]
- Power Features (Optional): [Any Integer Greater Than 1]

FIGURE 3.6.2.1 Feature Function Specifications
(Courtesy of Rapid Clip Neural Systems, Inc.)

ment Exclusion List" are not represented. If the "User-Supplied Subroutines" option is chosen, feature function subroutines should be attached to the client output file (section 3.1.3), according to *Rapid Learner*™ software specifications [1]. Each feature function subroutine receives all input measurement values one record at a time and converts them to the specified output "Number of Features."

"Historical Features" and "Power Features" are computed as functions of input features and user-supplied features (see section 6.2, 6.3). The "Number of Historical Steps" and "Historical Step Size" options specify the historical time points that historical features utilize. The most recent historical time point is always only one time point in the past. If more than one historical step is specified, then each step is separated by the "Historical Step Size" number of time points.

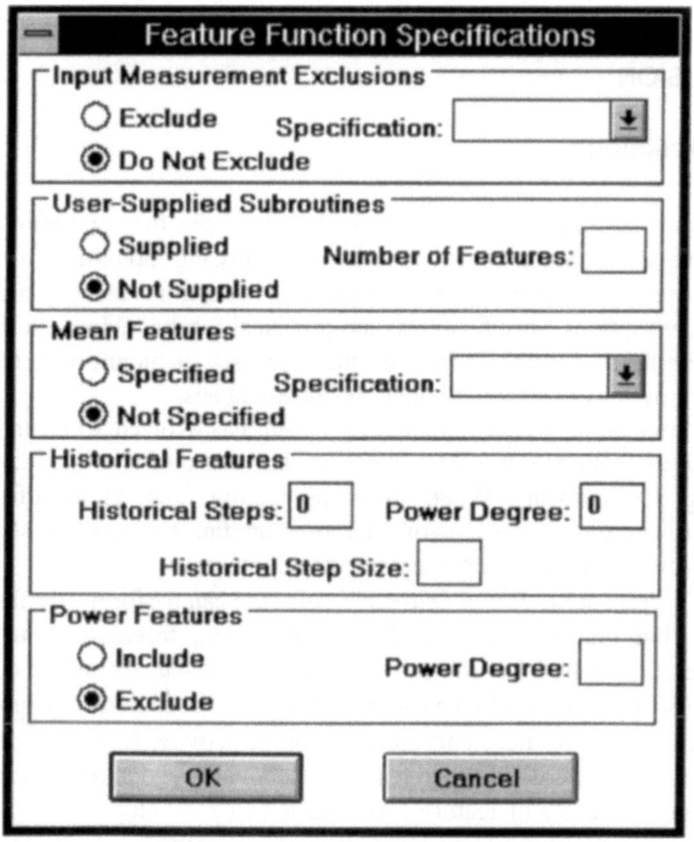

FIGURE 3.6.2.2 Feature Function Specification Display
(Courtesy of Rapid Clip Neural Systems, Inc.)

For example, if 3 is the "Number of Historical Steps," and 7 is the "Historical Step Size," then the input time points for recent feature evaluations at time point 100 are 99, 92, and 85. "Power Features" specifies the degree of polynomials that are correlated with dependent variables during prediction (see section 6.2).

Figure 3.6.2.2 shows the *Rapid Learner*™ dialog box for feature function specifications. The specifications are set at their default values, as follows:

- "Do Not Exclude" is selected under "Input Measurement Exclusions" — if "Exclude" were checked, feature numbers in the list would be excluded.
- "Not Supplied" is selected under "User-Supplied Subroutines."
- "Historical Steps" are set to 0 under "Historical Features"
- "Exclude" is selected under "Power Features."

CONCLUSION

Future Directions

The software system introduced in this chapter is the first operational version among several that are expected to evolve. Following modern software development trends, one planned growth direction is toward increasingly user friendly software descriptions, specifications, and graphical results. More substantive directions for future growth include an increased set of monitoring, forecasting, and control options to satisfy broader application needs.

The rapid learning neuro-computing system has evolved so far in ways that are similar to biological learning evolution, but are distinct from other neuro-computing developments. While automatic feature function identification has been the driving force for alternative neuro-computing systems (see section 4.3), fast adaptability has driven rapid learning development.

In its current state, *Rapid Learner*™ software can perform very fast learning operations but only during very simple monitoring, forecasting, and control. These operations parallel crude paired-associate learning operations during simple stimulus-response activity (see section 4.3). Just as biological neural systems evolved from crude to more interesting cognitive activity, *Rapid Learner*™ software is evolving toward expanded feature functionality and automatic refinement. For example, a variety of feature functions is being developed for visual image processing and related visual quality inspection applications. These feature functions will enable solutions to be developed for interesting problems, such as face recognition.

Interestingly, these software development directions are being driven by practical needs to solve particular problems, rather than academic desires to solve general problems. Even so, clear academic as well as practical benefits will result from developing more general rapid learning software and models, and practical as well as academic interest in developing them will be essential.

Summary

This chapter describes monitoring, forecasting, and control operations that are performed by the rapid learning neuro-computing system. It begins with an overview of input-output operation for a variety of system configurations, including single-computer systems, dual-processor systems, and multiple-computer client-server networks. The chapter follows with a detailed overview of *Rapid Learner*™ software operation. The overview includes examples of graphical user interface operation, input specifications, learning options, monitoring operations, forecasting options, control specifications, and feature function specifications. Interesting future directions include increasingly user friendly software development, increased software functionality from a user viewpoint, and increased software functionality from a cognitive modeling viewpoint.

REFERENCE

1. "*Rapid Learner*™ Batch Simulator User Manual," *RCNS Technical Report Series,* No. RLSIM97-04, Rapid Clip Neural Systems, Inc., Atlanta, GA, 1997.

4

Rapid Learning Models

INTRODUCTION . . . 100
4.1. RAPID LEARNING NEURO-COMPUTING SYSTEM OVERVIEW . . . 100
4.1.1. Modular Structure . . . 101
4.1.2. Kernel Hardware Module Overview . . . 106
4.1.3. Transducer Hardware Module Overview . . . 108
4.1.4. Manager Hardware Module Overview . . . 108
4.2. A RAPID LEARNING NEURAL NETWORK MODEL . . . 109
4.2.1. Component Neural Networks . . . 109
4.2.2. Transducer Neural Network Overview . . . 111
4.2.3. Kernel Neural Network Overview . . . 113
4.2.4. Manager Neural Network Overview . . . 115
4.3. A CONCURRENT BEHAVIOR AND MENTAL PROCESS MODEL . . . 115
4.3.1. Stimulus-Response Activity . . . 116
4.3.2. Mental Process Activity . . . 117
4.3.3. Behavior and Mental Process Coordination . . . 119
4.3.4. Behavioral and Cognitive Counterparts . . . 120
4.4. A RAPID STATISTICAL ESTIMATION MODEL . . . 122
4.4.1. Rapid Estimation Model Overview . . . 122
4.4.2. Ties to Linear Regression . . . 123
4.4.3. Ties to General Linear Models . . . 123
4.4.4. Ties to Non-linear Models . . . 123
4.4.5. Ties to Binary and Categorical Models . . . 126
4.4.6. Ties to Optimal Estimation . . . 127
CONCLUSION . . . 128
REFERENCES . . . 129

INTRODUCTION

Chapter 4 ties the rapid learning neuro-computing system to related neural network, mental process, and statistical estimation models. Section 4.1 presents key rapid learning neuro-computing system components. Section 4.2 describes a corresponding biological neural network model, section 4.3 describes a closely related mental process model, and section 4.4 describes a corresponding statistical model.

The models are presented in this chapter after the discussion of the operational computing system, as if the system was developed before the models. However, all key features of the computer system were inspired and shaped by previously developed models of computing, neuro-physiology, cognition, and statistical science. In particular, the rapid learning computing system was based on the following key concepts:

- Computer science explanations of coordinated multiple processor activity
- Neuro-biological science explanations of parallel neural network activity
- Cognitive science explanations of learning, behavior, and thinking
- Statistical science explanations of precise estimation and prediction

Science and technology fuel each other in useful ways [1]. The rapid learning technology described in this book has been created mainly to solve practical problems. Yet without inspiration from each scientific field that is represented in this chapter, the system could not have been conceived. On the other hand, as the rapid learning system expands to solve more practical problems in more interesting ways, it may point toward new scientific models within these fields as well.

4.1 RAPID LEARNING NEURO-COMPUTING SYSTEM OVERVIEW

This section introduces the underlying model for rapid learning software and hardware systems. The section begins with an overview of software and hardware modular structure (see section 4.1.1). The section follows with an overview of the main rapid learning system modules, including the following: the Kernel module that is the processing heart of the system (see section 4.1.2); the Transducer module that acts as a model specification interface between measurements and the Kernel module (see section 4.1.3); and the Manager module, which governs higher level operations including learning control and model refinement (see section 4.1.4).

4.1.1 Modular Structure

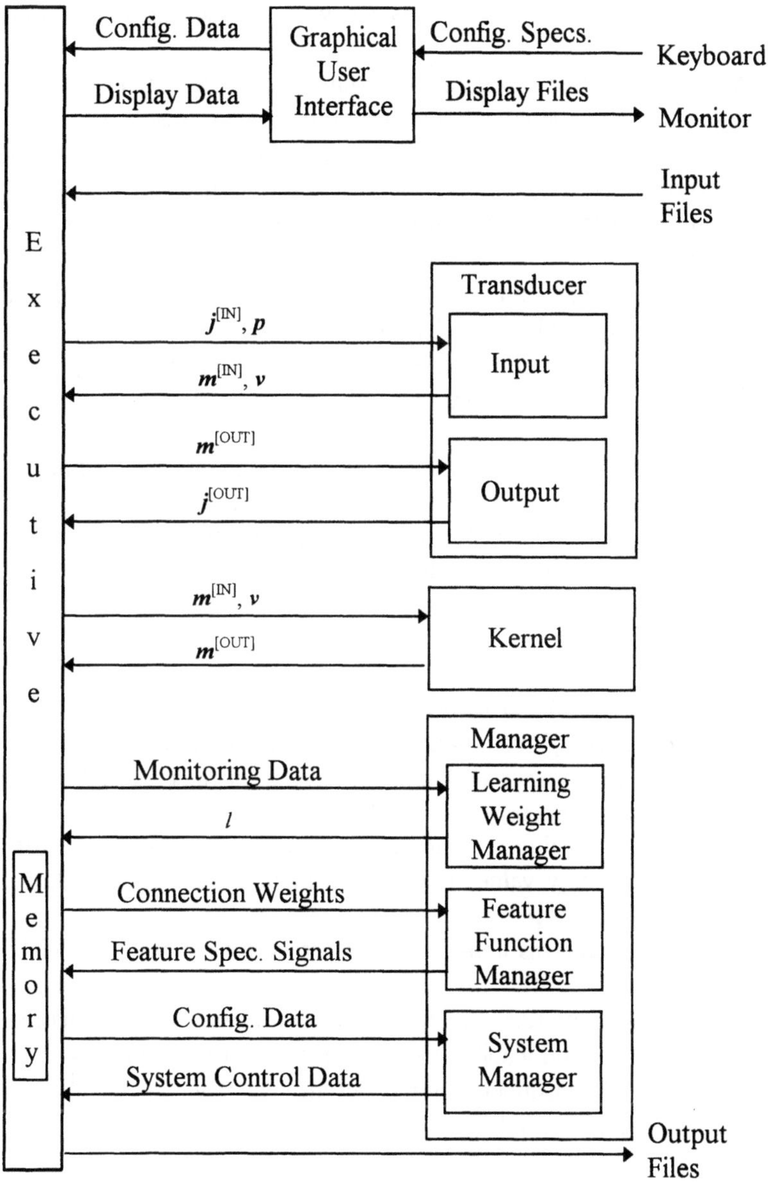

FIGURE 4.1.1.1 *Rapid Learner*™ Software Modular Structure (Courtesy of Rapid Clip Neural Systems, Inc.)

Figure 4.1.1.1 shows *Rapid Learner*™ software modular structure. (The following discussion is closely tied to the figure — readers are advised to have a copy of it handy for frequent reference.) Key software modules include the following:

- A Graphical User Interface that converts user commands to system configuration data and converts system output data to displays for the user.
- A Transducer that converts input measurements to input features and converts predicted features to predicted output measurements.
- A Kernel that performs core learning and prediction operations very quickly.
- A Manager that controls learning weight schedules, feature function modifications and high-level system functioning.
- An Executive that coordinates module and memory activity, much like a main program in a conventional software application.

Prior to concurrent operation, the Graphical User Interface supplies initializing and configuration data to the Executive and the Manager (see chapter 3). The Executive then assigns necessary memory for concurrent operation, while the Manager establishes initial learning weights and feature functions.

During concurrent operation, the Executive coordinates the following operations at each time point:

1. The Input Transducer receives input measurement values $\boldsymbol{j}^{[IN]}$ and plausibility values $\boldsymbol{p}$. (Throughout this book, bold case characters denote vectors and other arrays that contain several distinct variables — see the Glossary notation entry.)
2. The Input Transducer computes and returns non-missing feature values $\boldsymbol{m}^{[IN]}$ and viability values $\boldsymbol{v}$, based on input measurement values $\boldsymbol{j}^{[IN]}$ and plausibility values $\boldsymbol{p}$.
3. The Kernel receives non-missing feature values in $\boldsymbol{m}^{[IN]}$ and viability values $\boldsymbol{v}$.
4. The Kernel computes missing feature values based on non-missing feature values in $\boldsymbol{m}^{[IN]}$ and viability values $\boldsymbol{v}$, and it returns them in $\boldsymbol{m}^{[OUT]}$.
5. The Output Transducer converts output feature values $\boldsymbol{m}^{[OUT]}$ to predicted measurement values $\boldsymbol{j}^{[OUT]}$.

Along with these five concurrent operations that the Executive controls, the Manager controls feature functions and learning schedules that are used from trial to trial. In addition, the Manager governs occasional model refinement operations. Model refinement sometimes requires new feature function prediction, which may require concurrent operation interruptions. All refinement operations are governed by the Manager and coordinated by the Executive.

A monitoring application will be used in this section to outline *Rapid Learner*™ software operation. Monitoring instrument gauges requires prediction, deviance computation, and learning updating at each time point (see section 1.1). Before concurrent operation begins, the user supplies configuration specifications, such as measurement details, learning weight requirements, and monitoring requirements (see section 3.1). These specifications are converted to configuration data by the Graphical User interface module and supplied to the Executive. Accordingly, the Executive allocates memory and linkages to *Rapid Learner*™ modules for meeting these specifications prior to concurrent operation.

In a typical monitoring application, input measurement readings are supplied regularly in the form of a measurement vector $\boldsymbol{j}^{[IN]}$ containing one value of each measurement at each time point. Input measurements are converted to feature values by the Transducer, depending on application needs. For instance, a typical monitoring application requires that each measurement be predicted as a function of all other measurements at that time. It may also require a recent history of measurements for prediction. The Transducer computes feature functions of these historical measurements, in keeping with user configuration specifications and feature function control specifications.

For example, suppose that each of 250 gauges is being predicted as a weighted sum of the other 249 concurrent gauge values, along with each of the 250 gauge values during the previous 10 time points. In that case, the Transducer creates a feature vector $\boldsymbol{m}^{[IN]}$ at each time point by augmenting the current measurement vector $\boldsymbol{j}^{[IN]}$ with a history of measurement values going back 10 time points. The resulting feature vector $^{[IN]}$ at each time point has 2,750 elements [$2{,}750 = 250 \times (1 + 10)$].

In many practical settings, input measurements occasionally may be missing at different time points. In order to perform monitoring, forecasting and control operations in these settings, *Rapid Learner*™ software is designed to replace missing input measurements with predicted measurement values. The general mechanism for specifying missing values at each time point is a plausibility vector $\boldsymbol{p}$. The plausibility vector has as many elements as the input measurement vector $\boldsymbol{j}^{[IN]}$. If an input measurement is not missing, its corresponding plausibility value is 1; but if it is missing, its corresponding plausibility value is 0.

When the Transducer computes feature functions of measurement values that may be missing, it must also determine how to treat resulting feature values that may be completely or partially missing. The mechanism for establishing missing value status of each feature value is a viability vector $\boldsymbol{v}$. Just as the plausibility vector has one element for each measurement value, the viability vector has one element for each feature value. Also, if an input feature value is not missing, its corresponding viability value is 1; but if it is missing, its corresponding viability value is 0.

Monitoring and learning are automatically coordinated through internal *Rapid Learner*™ software operation. Coordinated monitoring operation is designed to be very fast through efficient use of the Kernel module. The Kernel module, in turn, is designed to perform the following operations very quickly:

- It receives an input feature vector $\boldsymbol{m}^{[IN]}$ from the Executive.
- It receives a viability vector $\boldsymbol{v}$ that indicates which elements of $\boldsymbol{m}^{[IN]}$ are missing and which are not missing.
- It receives an input learning weight l indicating how much impact the current input feature vector will have on learning.
- It uses the non-missing elements in $\boldsymbol{m}^{[IN]}$, along with previously learned connection weights to predict and store missing elements in $\boldsymbol{m}$.
- It returns the completed vector $\boldsymbol{m}^{[OUT]}$ to the Executive.
- It uses the non-missing elements in $\boldsymbol{m}^{[IN]}$, along with previously learned connection weights to create updated connection weights.

Monitoring at each time point can be initiated by calling the Kernel module once for each measurement. For the first call, the plausibility vector contains a 0 followed by all 1s, producing a predicted value of the first measurement from all others. Once that first predicted value has been supplied by the Kernel, it can be compared to the observed first measurement value and a monitoring deviance value can be computed. Likewise, each other measurement can be predicted in consecutive Kernel calls, once for each measurement. These multiple Kernel calls are coordinated by the Executive and governed by the Manager. (In practice, *Rapid Learner*™ software uses a modified Kernel that requires only one monitoring call.)

Besides coordinating Kernel calls, the Manager also coordinates general learning activity through the Learning Weight Manager. The Learning Weight Manager establishes learning weight schedules and determines when deviant measurements should not influence learning (see section 4.1.4). Learning weight schedules can be established so that all prior time points will have equal impact or more recent time points will have relatively high impact. Learning can also be interrupted if a measurement vector exceeds user-specified deviance criteria, or it can be scheduled to satisfy a variety of other adaptive processing needs (see chapter 5).

The Feature Function Manager also coordinates refinement activity (see section 4.5). During refinement, feature function connection weights that have been learned by the Kernel are examined by the Feature Function Manager. The Feature Function Manager identifies redundant features that can be combined or unnecessary features that can be removed. The Feature Function Manager then modifies feature function specifications accordingly. For example, if 250 measurements at each time point are being monitored as a function of all measurements during the last 10 points, all of the resulting 2,750 features (see above) may not be necessary. The Feature Function Manager is equipped to monitor connection weights among all 2,750 feature values occasionally and remove unnecessary features accordingly.

In some applications, predicted measurement values may be useful not only for establishing monitoring deviance but also for direct output. For example, suppose that a monitoring setting involves many correlated measurements, each of which can accurately be predicted from all the others. If a gauge breaks in that setting, output predicted values for that gauge will give accurate estimates for its values. Using such predicted values in place of observed values may allow processes to continue operating until a convenient shutdown period can be chosen to repair the broken gauge.

The path for predicted measurement output begins at the Kernel module, where predicted feature values $\boldsymbol{m}^{[OUT]}$ are available at the end of each time point. These predicted feature values are then converted to predicted measurement values $\boldsymbol{j}^{[OUT]}$ by the Output Transducer. The predicted measurement values are then sent to output files, as shown in Figure 4.1.1.1.

For some applications, Output Transducer functions are simple. For example, suppose that input features $\boldsymbol{m}^{[IN]}$ involve only input measurements $\boldsymbol{j}^{[IN]}$ augmented by historical measurement functions. Further suppose that no input values are missing. In that case, the Output Transducer returns values $\boldsymbol{j}^{[OUT]}$ that are simply predicted measurement values from the larger feature vector $\boldsymbol{m}^{[OUT]}$. In other applications, Transducer functions are more complicated, especially those involving many terms and partially missing input measurements (see section 5.3.5). In addition, the Transducer may compute output values other than predicted measurements. For example, it could compute deviance statistics for monitoring and control.

Figure 4.1.1.2 shows rapid learning <u>hardware</u> modular structure. The thick lines in the figure show input and output measurement and feature paths, which are designed for very fast through-put. The thin lines to and from the Manager board show feature function and learning weight control paths, which are designed for occasional rather than concurrent information processing. Unlike its software counterpart in Figure 4.1.1.1, rapid learning hardware utilizes several separate processors at the same time. The Transducer Board (or computer) may have a processor for computing each Transducer function at the same time. The Transducer Board may also have its own <u>Recent Feature Memory</u> (RFM) for operations such as monitoring, where functions of recent measurements are regularly used (see section 7.1).

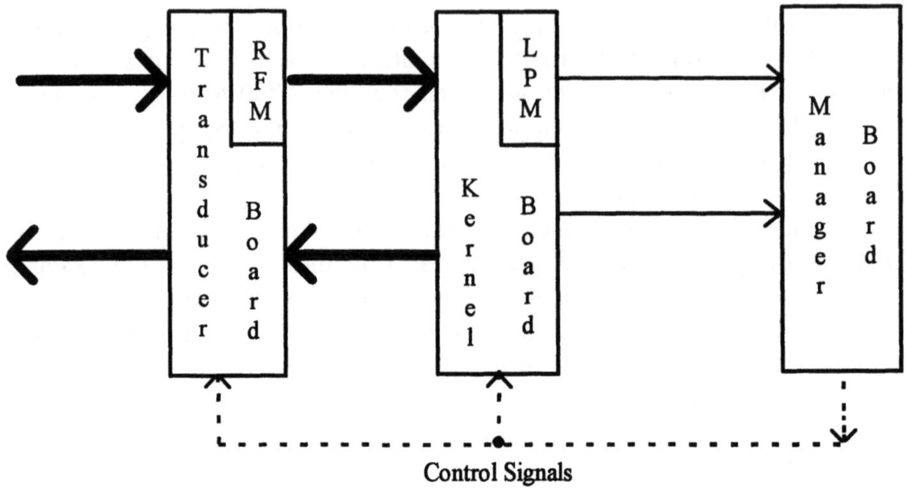

FIGURE 4.1.1.2 Rapid Learning Hardware Modular Structure
(Courtesy of Rapid Clip Neural Systems, Inc.)

The Kernel Board shown in Figure 4.1.1.2 includes a dedicated Learned Parameter Memory (LPM), which stores connection weights and other learned parameters. The Kernel Board may contain several Kernel hardware modules for forecasting, pattern completion, and other operations that may be separated into component Kernel operations (see section 6.4.2). Each Kernel module may also contain multiple processors (see below).

4.1.2 Kernel Hardware Module Overview

The key to concurrent operation is very fast prediction and learning updating, which may be achieved by parallel processor structures shown in Figure 4.1.2. The figure shows the top view of a rapid learning Kernel chip. This model depicts a layout for Kernel operation with 16 features, using 16 parallel processing arithmetic logic units (ALUs). Shown above the ALUs in the figure is a joint access memory (JAM), which includes nodes and busses. Nodes, which include memory elements and connection switches, are designated by circles. Busses are designated by horizontal and vertical lines.

Each ALU performs all operations necessary for predicting each feature that is dedicated to it [2]. In addition, each ALU participates in learning. The feature values and connection weights that are necessary for prediction and learning are provided through JAM nodes and switches.

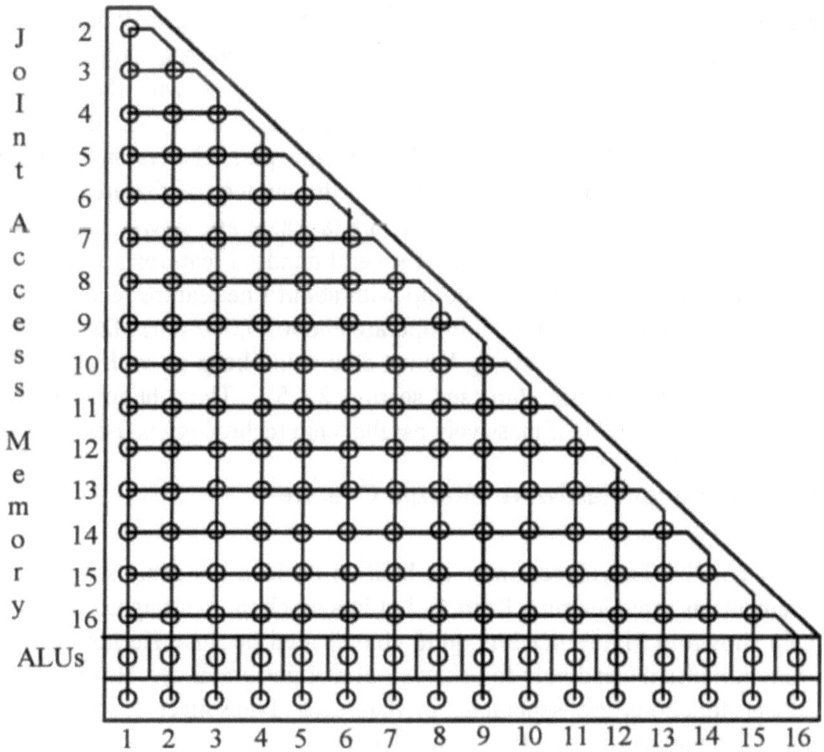

FIGURE 4.1.2 Kernel Module Architecture
(Courtesy of Rapid Clip Neural Systems, Inc.)

Predicting each feature value requires computing weighted sums of other feature values (see section 5.3). For example, predicting feature 1 from all others is possible through connections and memory nodes along the first JAM column shown in Figure 4.1.2. The nodes in this column contain connection weights for predicting the first feature value from all other feature values. The first ALU computes all the weighted sums for feature 1 by fetching these connection weights from nodes along its bus. The first ALU also fetches other feature values from their busses through nodes, connecting its bus to theirs.

Linkages from the first feature ALU to other ALUs involve the first feature bus, each other feature bus, and switches in the nodes between them. For example, the feature 2 value in Figure 3.1.2 is sent upward from the feature 2 ALU to the top of the chip, where it is then sent left to the top node on the feature 1 bus. It is switched through that node to the feature 1 bus and sent downward to the feature 1 ALU.

Kernel modules resembling the one shown in Figure 4.1.2 may be built using digital or analog chip technology [2-3]. While conventional, single processor software performs Kernel operations at rates that increase with the number of features squared, the digital parallel version operates at rates that increase only linearly with the number of features. The analog parallel version operates at constant rates that do not increase at all with the number of features. Accordingly, analog processing rates are fastest, digital rates are slower, and conventional rates are slowest. For example, if several hundred features are involved, a conventional Kernel module can keep up with about one feature record per second, while a digital Kernel chip could operate about 20,000 times faster than the conventional version, and an analog Kernel chip could keep up with about 1,000 times faster than the digital chip (see section 2.4.5). Thus, huge increases in learning speed are possible if massively parallel chip technology is used.

4.1.3 Transducer Hardware Module Overview

Massively parallel Transducers may be built to achieve the same kind of speed improvements as rapid learning Kernels, but in a much more straightforward way. The chip shown in Figure 4.1.2 is quite complicated because Kernel processors must function interdependently to achieve optimal performance. However, many Transducer functions are separable, in that each Transducer function can be computed independently of other Transducer functions. As a result, separate digital or analog processors can be dedicated to computing each Transducer function without difficulty. For example, spatial visual features in a pixel array can all be processed at the same time (see section 6.4.2). A variety of available digital and analog processors may be used to compute many related parallel Transducer functions [4-6].

4.1.4 Manager Hardware Module Overview

As with Kernel and Transducer operations, many Manager operations can be implemented using massively parallel processors. For example, the Manager refines and controls functions by comparing joint access memory elements copied from the Kernel. Most of the comparisons can be separated into component operations, which can be performed at the same time by parallel processors.

The Manager can perform occasional refinement operations while other concurrent operations proceed, with dead time occurring only during the transfer of connection weights from the Kernel LPM to the Manager. This dead time can be considerable if many features are involved and conventional computers are used.

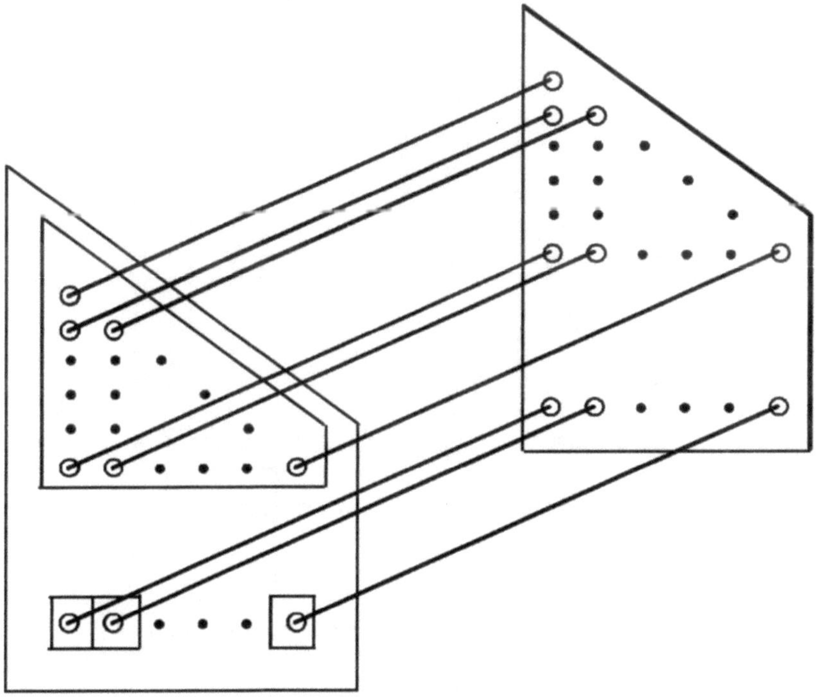

FIGURE 4.1.4 Kernel-Manager Communication
(Courtesy of Rapid Clip Neural Systems, Inc.)

Special-purpose hardware can be built, however, with massively parallel communication bus structures as shown in Figure 4.1.4. This kind of communication, using thousands of wires between a Kernel chip and a corresponding Manager chip is becoming achievable with modern multi-chip stacking technology [6].

4.2 A RAPID LEARNING NEURAL NETWORK MODEL

This section introduces a neural network counterpart to the rapid learning neurocomputing model. The section begins with an overview of neural network structure (see section 4.2.1). Neural network counterparts include a Transducer neural network (see section 4.2.2), a Kernel neural network (see section 4.2.3), and a Manager neural network (see section 4.2.4).

4.2.1 Component Neural Networks

Biological neural networks are marked by processing elements called neurons, interconnections between neurons called <u>synapses</u>, and hormonal mechanisms for

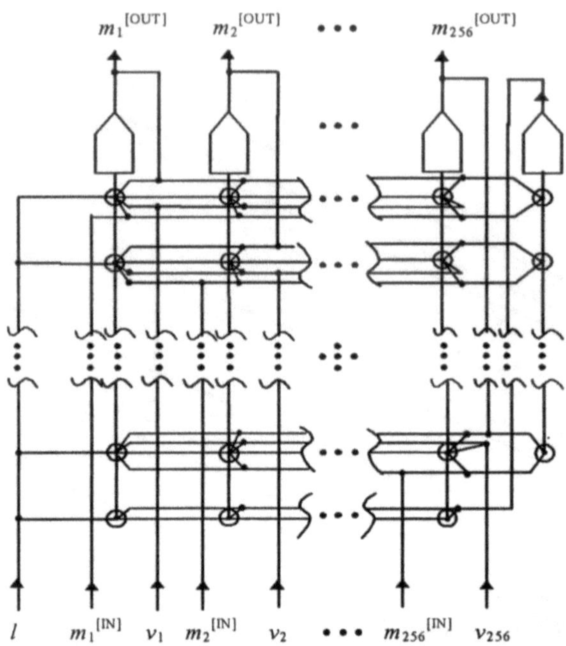

FIGURE 4.2.1 A Biological Neural Network Model
(Courtesy of Rapid Clip Neural Systems, Inc.)

modifying synaptic interconnection strength [7]. Figure 4.2.1 shows a model for a rapid learning biological neural network containing several neurons, indicated by the large pentagons at the top. Synapses are indicated by circles connecting the neuron output lines (axons) to neuron input lines (dendrites). Vertical lines at the bottom of the figure are input lines. Each neuron receives two inputs: its feature value $m^{[IN]}$ and its viability value v. In addition, each neuron receives outputs from all other neurons through synapses along its dendrite. An additional input line, connected to all synapses, represents a single learning weight l that influences all interconnections between neurons in the same way.

Each neuron also has one output measurement line $m^{[OUT]}$. Along with input-output neurons, the figure shows another neuron at the right. This neuron represents global information processing that can be utilized by all input-output neurons in the network.

The Figure 4.2.1 network shares some properties with alternative neural network models, such as Hopfield Networks [8-9]. However, the network has unique properties that are designed solely for real-time learning. For example, it differs from well-known multi-layer perceptron models (see section 2.3), in the following ways:

- Each neuron receives a concurrent learning weight for each input record.
- Each neuron receives an input viability value as well as an input feature value.
- Each neuron can supply either its input value or its predicted value as an output value.
- Each neuron receives inputs from all other neuron outputs as well as its own.
- Each neuron updates its learned prediction function before the next input record arrives.

The above differences give the rapid learning model added functionality relative to multi-layer perceptron models (see section 2.3). Provisions for learning weights, along with feedback from each neuron output to each neuron input, allow concurrent learning to occur (see section 4.3.1). Also, the input plausibility provision, along with input and output lines for each neuron, allow different measurements and features to play different input-output roles from trial to trial (see section 4.3.2).

Special cases of the Figure 4.2.1 neural network model perform Transducer, Kernel and Manager functions (see section 4.1). Connections from all outputs to all inputs are not necessary for some special cases, such as Transducer networks (see section 4.1.2). Symmetric connections between pairs of neurons are necessary for other special cases such as Kernel networks, rather than two distinct connections from each member of a given pair to the other member (see section 4.1.2). Other networks that serve only interconnection roles do not require learning provisions (see section 4.1.4).

4.2.2 Transducer Neural Network Overview

Figure 4.2.2 shows a Transducer input-output example, which has been studied for rapid learning during speech recognition [10]. The inputs are arranged by rows and columns. Each row corresponds to a frequency band that might be received from a region of the cochlea in the ear [7]. Each column corresponds to a time slice, ranging from most recent in the right column to least recent in the left column. In cognitive modeling terms, the columns might be stored in a short-term cache [11]; in *Rapid Learner*™ terms the columns might be stored in RFM (see section 4.1).

The Transducer converts input measurements to three output feature values, each of which is a third-order function of the inputs. The left-most output has three input lines. Its lower input line has a synapse for each individual input. The lower line thus represents a weighted sum of each individual input, with the

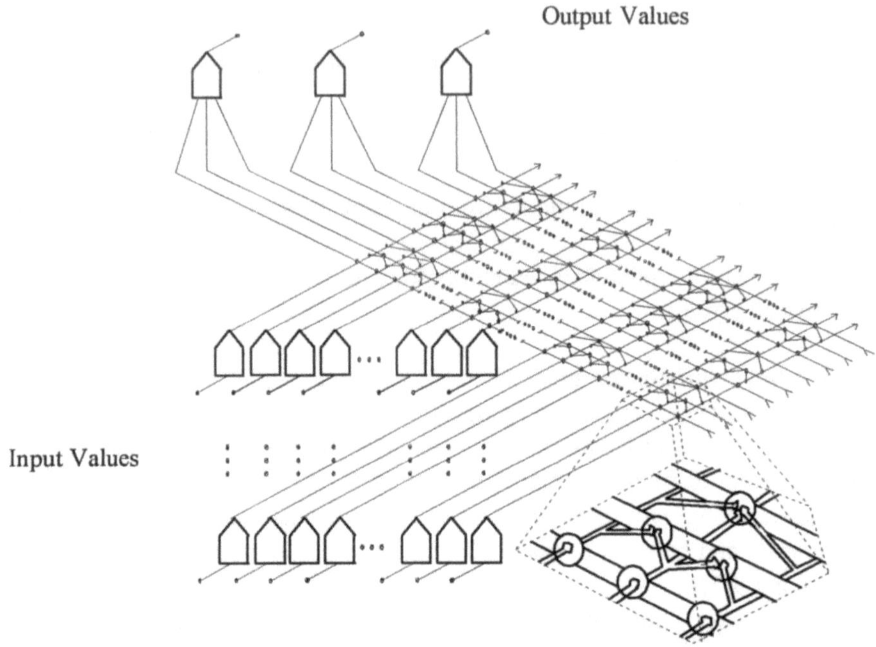

FIGURE 4.2.2 A Transducer Input-Output Example
(Courtesy of Rapid Clip Neural Systems, Inc.)

weights being stored in the connecting synapses. Its middle input line has synapses connected to individual input pairs. Each of these synapses multiplies the input pairs and places the product on the output line. Likewise, its upper input line represents a weighted sum of input three-way products.

The Figure 4.2.2 Transducer represents three output functions, each of which is a third-order function of the input functions. These three functions may each represent auditory features, which may be directly used for distinguishing phonemes from each other [10]. The three functions may also be used as inputs to a Kernel neural network, which can learn how they should be combined for optimal speech recognition. Three corresponding viability functions may also be used in conjunction with Transducer functions, to allow coordinated mental process activity in which auditory patterns can be completed as well as recognized (see section 4.3).

The Transducer network differs from Kernel neural networks (see section 4.2.3), in that it has no provision for learning. The Transducer could be based on prior learning, however. The Transducer might be configured by a Manager neural network (see section 4.2.4) after a third-order model has been learned by a

Kernel neural network. Alternatively, the Transducer might be hard-wired into a biological neural system prior to learning initiation.

It should be noted that the Transducer produces a certain *non-linear* function of input measurements, simply by combining synaptic inputs multiplicatively. Furthermore, simple Transducer variations can produce far more general non-linear functions. For example, a Transducer can be produced with one output for each input, one output for each pair-wise product of inputs, and so on. In this way, weights can be assigned to (or learned for) each synapse such that *any* input-output function can be approximated (or learned) to *any* degree of accuracy [10].

Thus, neural network systems that use multiplicative Transducers can represent practically any input-output function, subject only to physical size and interconnection limitations. In the same way *Rapid Learner*™ networks can learn and use a broad variety of linear and non-linear relationships (see section 6.3-6.5). Finally and most importantly, the Transducer network, in conjunction with its companion Kernel neural network (see below), shares a key property with other biological neural network models: the capacity to predict and learn non-linear relationships *in real time*.

4.2.3 Kernel Neural Network Overview

Figure 4.2.3 shows a rapid Kernel learning neural network model. The Kernel neural network closely resembles its counterpart in Figure 4.7.1. Each output feature value, shown by arrows at the top of the figure, comes from a dedicated neuron. Each neuron has its own input viability line and input feature line. The input learning weight line, shown at the lower left, is connected to all neurons in the network, just as its computing system counterpart. Also, the array of synapses, thin axon lines, and thick dendrite lines at the top of the figure closely resemble their Joint Access Memory counterparts in Figure 4.1.2. An additional neuron, shown crossing all feature neurons horizontally at the bottom, performs network-wide operations. It also supplies an output value, shown at the right, which is a global measure of input feature vector deviance (see section 5.1).

Just as its neuro-computing counterpart, the Kernel neural network may be viewed as a fast engine that enables several mental activities to occur concurrently and quickly. In this case, activities can be grouped into mental activity categories such as stimulus, response, learning, and thinking (see section 3.3). While electronic components can convert inputs to outputs in the nanosecond range, neurons only convert inputs to outputs in the millisecond range [7]. The Kernel neural network, however, can convert inputs to outputs very quickly because each neuron can process information at once and all neurons are highly interconnected. In particular, the figure represents a neural network process that converts all input feature values to output feature values within the time of one

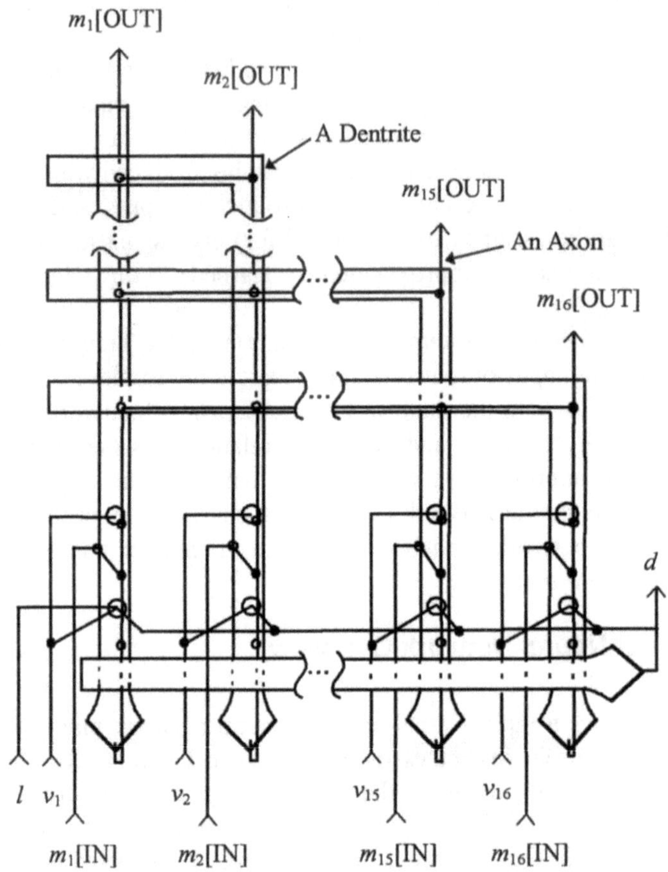

FIGURE 4.2.3 Kernel Neural Network Layout
(Courtesy of Rapid Clip Neural Systems, Inc.)

neuron firing cycle, and updates all synapse learning within one additional neuron firing cycle [3]. For this reason, the model may provide concrete explanations for rapid learning mechanisms that have been observed in biological experiments [12].

The Figure 4.2.3 Kernel by itself is a linear system, in that it predicts each feature value as a weighted sum of all other feature values. However, the system is non-linear when input features include non-linear functions of measurements. As a result, the Kernel and Transducer can combine to perform highly non-linear learning and prediction. Moreover, they can learn relationships among many variables very quickly — far more quickly than neural network systems that operate iteratively rather than recursively (see section 2.3).

4.2.4 Manager Neural Network Overview

Like its neuro-computing counterpart, the Manager neural network controls feature functions, learning weights, and overall neural system operation. In biological neural systems, many feature functions are consolidated at birth, such as basic structures for converting visual and auditory sensory information to frequency and edge detection features [13]. The Manager neural network functions by receiving such sensory input values and determining if and how they should be interconnected with motor output values for optimal biological control.

Like its neuro-computing counterpart, the Manager neural network refines input-output connections by examining connection weights between measurement features. Since the underlying mathematical operations for refinement are rather simple (see section 8.2), neural network architectures for refinement are simple as well. For example, suppose that the Manager is determining whether or not a pair of features is redundant, say features 1 and 2 in a feature vector $\boldsymbol{m}$. Suppose that the synaptic connections in Figure 4.2.3 are labeled as elements in an array $\boldsymbol{\omega}$. Suppose further that each connection is ω_{rc} for row r and column c. The redundancy assessment for feature 1 and feature 2 requires computing $|\omega_{32}-\omega_{31}|$, adding it to $|\omega_{42}-\omega_{41}|$, and so on. A single biological neuron that determines whether or not features 1 and 2 are redundant would thus have all connection weights involving features 1 and 2 as inputs. The neuron would create an electrochemical signal that reflects the sum of these difference magnitudes. The neuron would fire only if the sum of all such magnitudes were below a cutoff threshold. At the same time, other neurons could be assessing other feature pair redundancy levels. Likewise, other Manager refinement operations could be performed by other simple neural structures, operating in a highly parallel way.

4.3 A CONCURRENT BEHAVIOR AND MENTAL PROCESS MODEL

This section introduces a behavioral and cognitive counterpart to the rapid learning neuro-computing model. The section begins with a description of stimulus-response activity in rapid learning terms (see section 4.3.1), followed by a description of simple mental process activity in rapid learning terms (see section 4.3.2). The section follows with a description of Manager operations in mental process terms (see section 4.3.3). The section ends with an overview of more complicated behavioral and cognitive operations in rapid learning terms (see section 4.3.4).

4.3.1 Stimulus-Response Activity

Figure 4.3.1 shows input and output stimulus-response variables to a "Mental Process" model, which will be described later in this section. The inputs include a plausibility input variable ***p***, a learning ***reinforcement*** *l*, and an input-output *judgment* variable ***j***. A plausibility variable ***p***, learning reinforcement *l* and the input judgment variable ***j*** arrive at the beginning of each time point as indicated by the up arrows. At the end of each time point, the output judgment variable leaves the system as indicated by the down arrow.

The judgment variable ***j*** is a vector, some or all elements of which may be missing at the beginning of each time point. The plausibility variable ***p*** is also a vector with the same number of elements as ***j***. An element in ***p*** has a value of 1 if its corresponding element in ***j*** is missing, and it has a value of 0 if its corresponding element in ***j*** is not missing.

As with its neuro-computing counterpart, at each time point the Mental Process performs the following steps:

- It receives input judgment, plausibility, and learning weight values.
- It determines which input judgment values are missing from the input plausibility vector.
- It predicts — using prior learned parameter values — each missing judgment value as a function of judgment values that are not missing.

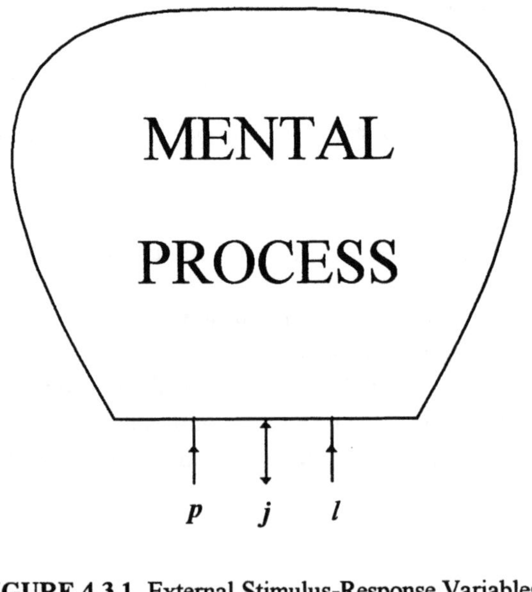

FIGURE 4.3.1 External Stimulus-Response Variables
(Courtesy of Rapid Clip Neural Systems, Inc.)

- It updates learned parameter values, using input measurements, plausibility values and the current learning weight.

Figure 4.3.1 can explain a broad variety of mental activity inputs and outputs. For example, each incoming judgment vector might contain spatial as well as auditory information. In that case, the task of announcing the name of a viewed object may be represented by a time trial with a plausibility vector containing 0s for auditory judgment vector elements and 1s for spatial judgment vector elements. Likewise, the task of drawing an object that has just been seen may be represented by a time trial with a plausibility vector containing 1s for auditory judgment vector elements and 0s for spatial judgment vector elements. In addition, the task of learning the name of an object may be represented by a plausibility vector containing all 1s in a time trial, along with a positive learning weight for the time trial. This kind of *paired-associate* learning and performance (see section 5.2) can be further expanded into an array of simple cognitive operations, all of which fall under the input-output structure shown in the figure.

4.3.2 Mental Process Activity

Figure 4.3.2 shows the basic modules within the concurrent learning and performance Mental Process model. They include the following:

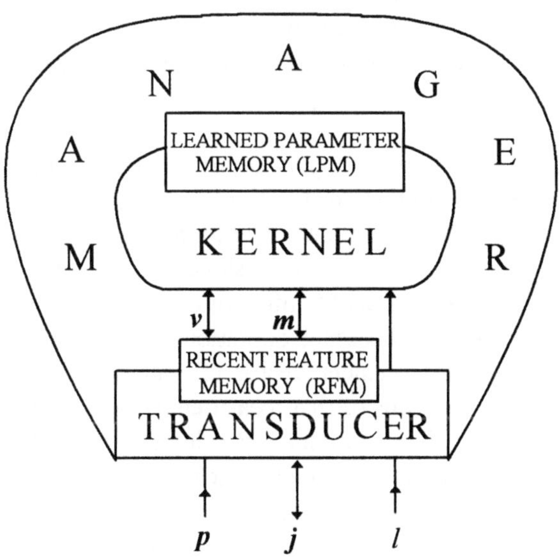

FIGURE 4.3.2 Mental Process Model Structure
(Courtesy of Rapid Clip Neural Systems, Inc.)

- A Transducer interface between external measurements and internal mental state variables.
- A Recent Feature Memory (RFM) that may be used by the Transducer to compute input mental function values.
- A Kernel module that performs concurrent operations very quickly.
- A Learned Parameter Memory (LPM) that contains connection weights among mental state variables.
- A Manager module that coordinates overall mental process activity.

As with its neuro-computing counterpart, at each time point the modules perform the following steps:

- The Transducer receives the input judgment vector $\boldsymbol{j}^{[IN]}$ and its corresponding plausibility vector $\boldsymbol{p}$.
- The Transducer converts input judgment and corresponding plausibility values to input *mental state* (synonymous with feature) values in $\boldsymbol{m}^{[IN]}$ and corresponding viability values in $\boldsymbol{v}$.
- The Transducer updates recent features in RFM.
- The Kernel predicts — using learned parameters stored in LPM — each missing judgment value as a function of judgment values that are not missing.
- The Kernel uses the input judgment values in $\boldsymbol{j}^{[IN]}$ to update LPM values according to the input learning reinforcement l.
- The Transducer converts predicted "mental state" values in $\boldsymbol{m}^{[OUT]}$ to predicted measurement values in $\boldsymbol{j}$.

Meanwhile, the Manager governs input-output timing and the Transducer functions that determine input-output relationships (see section 4.2.2) The Manager also occasionally refines Mental Process operation (see section 4.2.4).

The Mental Process model structure shown in Figure 4.3.2 has been designed to reflect all *Rapid Learner*™ software operations as well as a variety of simple human information processing operations. In terms of the figure, the Transducer receives a measurement vector at each time point, after which it computes mental state functions of that measurement vector and historical measurements stored in RFM. These functions are elements of the mental state vector $\boldsymbol{m}$ shown in the figure. The Kernel module then uses learned connections among mental state elements to predict each measurement for monitoring. The Kernel module also updates these learned connections before the next time point (see section 4.3.2). The Manager determines the functional relationship between measurements and mental state variables.

To give a human information processing example, the process of associating auditory and spatial information involves converting input sensory information (measurements) to perceptual features. For example, pixel information from sensory rods and cones in the retina of the eye is converted to edge, corner and curve features as it passes from the retina toward the central nervous system during visual perception [13]. Likewise, sound information received by the cochlea of the ear is transformed to frequency information as it is received. Thus, in Figure 4.3.2 terms, the cochlea of the ear and visual feature detection neurons perform input Transducer functions that transform input measurements to internal mental states.

Continuing the human information processing example, consider the process of drawing a picture of a cow after a person has been told, "Please draw a cow." The process involves first receiving the sound information and converting it to auditory features in the Transducer. At the same time, spatial information that is not received is missing information that must be predicted. The features associated with the sound "cow" are then supplied as inputs to a Kernel that has learned the required auditory and spatial connections. The Kernel then computes the required prediction functions and supplies them to the Transducer. Finally, the Transducer computes the required output functions and supplies them to the motor system.

Simple human learning is often described as the presentation of paired information at the same time as a reinforcement [11]. For example, a child who is congratulated for pointing to a cow when asked to do so receives a positive reinforcement. In terms of Figure 4.3.2, reinforced learning is represented by an input auditory-spatial vector with no missing elements, along with a positive learning reinforcement (l in the rapid learning neuro-computing model.) The reinforcement strength determines learning impact, in that a more substantial reinforcement produces a larger change in feature connection weight values.

4.3.3 Behavior and Mental Process Coordination

The Figure 4.3.2 model may represent complex mental processes in one of two ways: either simple versions of the model may be controlled by external processes, or complex versions of the model may coordinate complex operations internally. For example, in early developmental or lower evolutionary mental processes, external stimulus and reinforcement play larger roles in mental activity than internal mental process control plays. The Manager in this model may be viewed as being larger in more developed mental processes, thus resembling the cerebral cortex.

As with *Rapid Learner*™ software operation, the process of identifying individual measurements as missing or not missing at each time point can be performed externally, by supplying multiple plausibility vectors to the Figure 3.3.2

model. Alternatively, the process of predicting each measurement as a function of all other measurements can be controlled internally. The internal mode has several advantages. First, the internal process can be controlled automatically and without external effort or thought, just as it is by *Rapid Learner*™ monitoring software (see section 2.2). Second, the internal process reflects human capacity to monitor many sensory events at the same time, automatically. Finally, the parallel internal process reflects human capacity to monitor many sensory events at once, identifying those that are unusual very quickly.

Human information processing systems regularly refine internal prediction models as a natural part of thinking and dreaming [11]. In rapid learning terms, thinking may be viewed as a parallel computing process that may take place more slowly than — but at that same time as — automatic stimulus, response, and learning activity. Dreaming may be viewed as a stand-alone process during which all mental activity is devoted to model refinement.

Therefore, simple versions of the Figure 4.3.2 model, combined with externally coordinated stimulus and reinforcement, can explain simple mental activity; but far more powerful versions can explain more complicated and automated internal mental process activity. Indeed, while concurrent learning methods have been designed primarily for solving practical problems in real time, they may also be useful for describing related mental activities.

4.3.4 Behavioral and Cognitive Counterparts

Psychological theory is often separated into behavioral and cognitive schools [11]. Without attempting to explain internal thought processes, behaviorism focuses on stimulus, response, and learning relationships that have been observed from psychological experiments. In contrast, cognitive theory focuses on developing and testing theories of the process itself. The Mental Process Model shown in Figure 3.3.2 has been partly inspired by behavioral models. Most notably, the model reflects the basic behavioral model properties of concurrent paired-associate stimulus, response, and learning. The Mental Process Model also has a working, operational component, namely *Rapid Learner*™ software. As a result, the model has an operational explanatory cognitive side as well as a stimulus-response behavioral side.

The Figure 4.3.2 model shares structural as well as operational properties with established behavioral and cognitive counterparts. The model's explanatory role in stimulus, response and learning has already been outlined (see section 4.1.1). Simple monitoring and paired associate learning are the simplest among a broad variety of information processing operations that can be explained by the Mental Process model. The model also explains the human capacity to complete patterns that are partially missing (see section 6.3), in a way has stimulated so-called "Gestalt" theory and other theories of human perception [13].

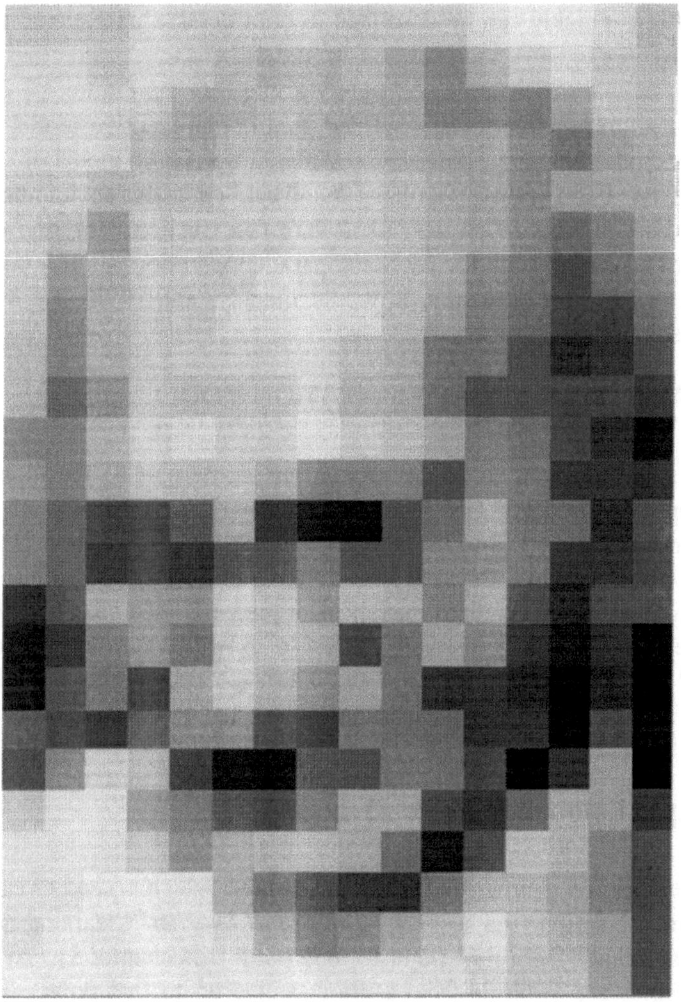

FIGURE 4.3.4 A Perceptual Imputing Example
(Courtesy of Rapid Clip Neural Systems, Inc.)

Figure 4.3.4 shows a higher-order perceptual process that the Mental Process model can explain. When viewed close up with eyes wide open, the figure looks like many meaningless, overlapping squares. When viewed at a distance through squinting eyes, however, the figure reveals the well-known likeness of a legendary Indian leader. In Mental Process Model terms, all input pixels are treated as non-missing and are assigned plausibility values of 1 when viewed close up with eyes wide open. When some of these plausibility values are set to 0, however, the mental process uses its capacity to quickly predict missing values in order to complete the picture. Furthermore, this mental process capacity to identify patterns quickly can be described in considerable operational detail (see section 6.3).

Finally, cognitive scientists will rightly argue that the Figure 4.3.2 model is very simple — too simple to be regarded as a comprehensive model of human thought. It should be noted, however, that this is a *working* model of *concurrent* mental activity. Each activity that has been described in this section has a working counterpart in *Rapid Learner*™ software (see section 3.1), and each such counterpart has the capacity for very fast, parallel operation. It may be simple as a cognitive model, but it works, it works quickly, and it has the capacity to evolve considerably (see section 8.2).

4.4 A RAPID STATISTICAL ESTIMATION MODEL

This section introduces a statistical estimation counterpart to the rapid learning neuro-computing model. The section begins with an overview of the rapid learning model in statistical terms (see section 4.4.1). The section follows with an overview of the rapid learning system in terms of linear regression (see section 4.4.2) as well as other standard statistical procedures (see section 4.4.3). The section then ties the rapid learning system to extended non-linear, binary, and categorical models (see sections 4.4.4 and 4.4.5; see also chapter 6). The section ends with an overview of rapid learning optimality characteristics in statistical terms (see section 4.4.6).

4.4.1 Rapid Estimation Model Overview

Statistical estimation is typically based on a sample, which is used to estimate parameters for prediction (see section 2.2). *Rapid Learner*™ learning is based on samples that change from time point to time point (see chapter 5). At each time point, estimates are available from a prior sample up to that point. These estimates are then used to predict missing feature values in $\boldsymbol{m}^{[IN]}$. Once prediction has been completed, estimates are updated, based on an updated sample that includes that time point.

Rapid Learner™ prediction speed is attained by special functions that have been designed to update parameter estimates recursively [2]. Recursive *Rapid Learner*™ estimates are updated very quickly, without the need to save an entire prior sample. These estimates are updated based on prior estimates and current measurements only, in such a way that the measurements need not be stored after the trial. These recursive estimates are related to statistical estimates, called *recursive least squares* estimates [14], which avoid time-consuming matrix inversion computations. However, *Rapid Learner*™ estimates are also uniquely tailored to suit concurrent operation and massively parallel processing needs.

Rapid Learner™ software also utilizes learning weights to produce estimates that may be differentially impacted by different measurements. These learning weights also operate recursively, in such a way that the entire prior learning weight history is not needed. Instead, each updated *Rapid Learner*™ estimate is based on a weighted combination of prior sample statistics and the current measurement record. Within the this estimation scheme, the learning weight governs how the prior sample and the current measurement record are combined to produce posterior estimates (see chapter 5). Such weighted learning is tied to an area of statistical estimation called Bayes decision theory (see section 4.4.6).

4.4.2 Ties to Linear Regression

Linear regression is one of the most widely studied and used statistical estimation models [15-17]. A recursive, weighted variant of linear regression is also the simplest special case of the Rapid Estimation Model. Like linear regression prediction estimates, Rapid Estimation Model predictors are weighted sums of other measurements (see section 5.2).

The linear regression version of the Rapid Estimation Model is simple and fast, relative to more general versions (see below). The linear regression version requires no Transducer because each measurement is predicted as a linear function of other measurements only. Also, linear Kernel operations are substantially faster than if additional historical, categorical, or other features are used.

4.4.3 Ties to General Linear Models

Explaining the distinction between linear regression models and general linear models requires a preliminary discussion of measurement types (see section 6.1, 6.5). Measurements can be classified as arithmetic, binary, or categorical [18]. Arithmetic measurements are distinguished by their additive (synonymous with linear) correlations, in that each among several arithmetic measurements is correlated with a weighted sum of the others. Arithmetic measurements are also distinguished by their value range, which is any negative or non-negative number. For example, human height and human weight are arithmetic variables, insofar as a line on a height versus weight plot describes their correlation. Binary measurements may also be linearly correlated with arithmetic or other binary variables, but their values are restricted to either 0 or 1. For example, items on a multiple choice test are binary if they are coded as such (0 for "Fail," 1 for "Pass")

A categorical variable has values that are arbitrarily assigned numerical labels for each distinct category, ranging from 1 to the maximum number of categories. Examples of categorical variables include "race" and "continent" if they are labeled categorically (e.g., 1 for "Caucasoid," 2 for "Mongoloid," 3 for "Negroid;" 1 for Asia, 2 for South America, etc.). Categorical variables cannot be linearly associated with each other or with other variables, because they have more than two values that are arbitrarily assigned. However, categorical variables may always be transformed to equivalent binary variables that may be linearly associated with other variables. When *Rapid Learner*™ operations involve categorical measurements, such transformations are automatically performed by the Transducer (see section 6.4).

Unlike linear regression models, general linear models deal with relationships among multiple variable types, some of which may be binary or categorical [18]. Like *Rapid Learner*™ estimation, most general linear model estimation techniques are based on converting categorical measurements to binary counterparts and then using linear regression to correlate them. Several standard general linear models and techniques are available for dealing with different variable types, including those listed in Figure 4.4.3.

Like the general linear model, the Rapid Estimation Model and its operational *Rapid Learner*™ counterpart can deal with arithmetic, binary and categorical measurements, both as independent variables and dependent variables. Moreover, *Rapid Learner*™ can deal with all possible combinations of variable types as independent and dependent variables, not just those listed in Figure 4.4.3. In addition, the software can treat any variable as either a non-missing independent variable or a missing dependent variable at any time point. For example, consider the case in which each measurement record includes a person's age,

Independent Variable Type(s)	Dependent Variable Type(s)	General Linear Model(s)
A	A	simple regression
As	A	multiple regression
As	As	multivariate regression
B	A	t-test
B	As	Hotelling's T^2
C	A	one-way analysis of variance
Bs, Cs	A	factorial analysis of variance
As, Bs, Cs	A	analysis of covariance
As, Bs, Cs	As	multivariate analysis of covariance
As	Bs, Cs	discriminant analysis

FIGURE 4.4.3 Rapid Learning Measurement Specifications for General Linear Models (A: arithmetic; B: binary; C: categorical; s: more than one — Courtesy of Rapid Clip Neural Systems, Inc.)

height and weight, treated as arithmetic variables; the person's disease history ("yes" or "no" for each item on a list), treated as binary variables; and the person's race and residential region, treated as categorical variables. If only the binary and categorical measurements are plausible for a given person, *Rapid Learner*™ operations would predict the person's arithmetic age, height and weight variables, as in a multivariate analysis of covariance. If only the arithmetic variables are plausible, then the software would predict the person's disease history status, race and residential region as in a discriminant analysis, and so on. By using plausibility values selectively in this way, *Rapid Learner*™ operations can subsume a broad variety of linear models, including all that fall under the general linear model category and others (see section 4.4.5).

Thus, the *Rapid Learner*™ counterpart to the general linear model is closely related, but it is more general in several ways. Other ties to and contrasts with general linear models are described later in this book (see section 5.5 and chapter 6).

4.4.4 Ties to Non-Linear Models

Rapid learning neuro-computing operations and related statistical models incorporate measurement non-linearities in several ways, including different uses of the following measurement properties:

1. Two non-linear functions of time may be linearly related to each other.
2. If a single measurement is a non-linear function of time, current values of the function can be accurately predicted by a linear combination of previous values.
3. Non-linear relationships among variables can be represented as linear correlations among non-linear functions of the same variables.

Figure 4.4.4 illustrates the first property for a strain gauge monitoring example that was presented earlier (see section 1.1.2). In that example, several strain gauges were monitored, each of which was related to time in a highly non-linear way, as shown in the figure. Yet, since each non-linear relationship was similar, the gauges were linearly correlated with each other. As the figure shows, if values for each of two such gauges are placed in the same plot, with one point for each measured time point, the resulting points fall along the same line.

To illustrate the second property, the missile tracking example presented earlier involves tracking an object that was moving along a highly non-linear path in space (see Figure 1.3.1). Yet, *Rapid Learner*™ software was able to track the object by treating each new location as a linear function of a few immediately preceding locations. This way of linearly linking non-linear time functions is not

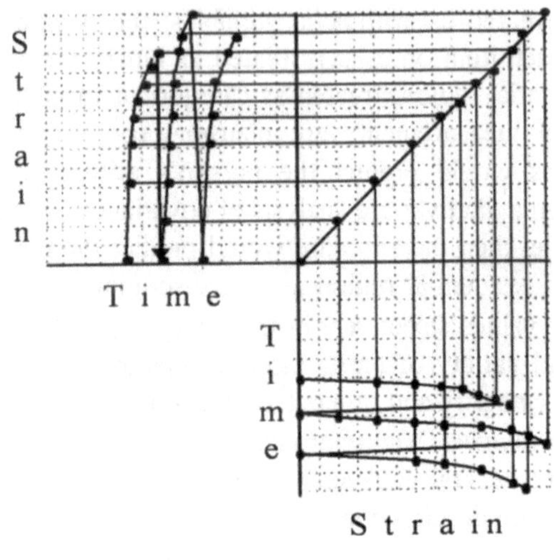

FIGURE 4.4.4 Linear Associations Among Two Non-linear Time Functions (Courtesy of Rapid Clip Neural Systems, Inc.)

new [20]. The novel aspect of such use by *Rapid Learner*™ software is its capacity to operate with these functions very quickly.

To illustrate the third property, the pattern recognition example presented earlier (see section 1.4) requires that multiplicative functions of pixel values in a visual array be used to distinguish among pattern types. For that example, second-order cross-products are required among the 49 pixels in each pattern (see section 1.4). *Rapid Learner*™ software represents such functions by computing cross-product features, in that case 1,176 of them, and *linearly* correlating them with other features. The key to being able to deal with such non-linear features effectively is the *Rapid Learner*™ software capacity to handle many such feature functions very quickly (see sections 2.4.1 and 2.4.2).

4.4.5 Ties to Binary and Categorical Models

Along with linear associations among arithmetic variables and others, *Rapid Learner*™ software can represent linear associations among purely binary and categorical variables. Related statistical estimation models have not been included under the general linear model category historically, but under a variety of alternative statistical models for dealing with binary and categorical data [21-23]. These include log-linear models and logistic regression models, among others. In addition, a closely related neural network model has been developed [24]. The model is equivalent to a rapid learning model that represents all possible powers among a set of binary measurements.

A related psychometric *conjunctive* model has been used to measure learning during testing in humans [25-26]. Rapid learning models, in contrast, have been developed for learning during information processing by machines. Along with a variety of technical similarities, conjunctive psychometric models share the key rapid learning property: the capacity to represent learning *during* information processing performance.

4.4.6 Ties to Optimal Estimation

Linear regression and related procedures are used with wide success because they produce optimally precise parameter estimates, based on standard random variable assumptions from statistical theory [15-20]. Regression estimates are optimal according to several criteria, including minimizing expected squared error [15]. Related linear rapid learning estimates share the same properties as regression estimates, because they are mathematically equivalent to linear regression estimates from prior samples (see section 5.2).

When non-linear measurement features are linearly correlated, rapid learning estimates are equivalent to estimates from a very general statistical model called the *exponential family* model [15,17]. Just as exponential family estimates have optimal asymptotic properties based on suitable assumptions [*ibid.*], rapid learning estimates share the same properties based on similar assumptions [18].

Rapid Learner™ software operations are also equivalent to Bayes statistical estimation procedures [27], which deal with cases where different relationships may exist among the same measurements from time to time. These close ties to Bayes estimates may be helpful in justifying *Rapid Learner*™ software results in non-stationary settings. Related rapid learning operations may also be justified for applications where subjective information must be combined with measurement observations to make decisions — an application where Bayes estimates have been widely studied and used in the past [18,26].

Along with optimal precision, *Rapid Learner*™ software procedures share one other key property with related statistical procedures: *identifiability* [17]. A model is called identifiable if no two distinct sets of estimates can explain data in precisely the same way. Identifiabile models have powerful estimation properties, including guarantees that estimates will not depend on extraneous concerns such as initial values. Many alternative neuro-computing models do not share such key properties, because they are not identifiable (see section 4.3).

CONCLUSION

Future Directions

Model development is sometimes viewed as an ongoing *alignment* process, during which some models are linked to other models and empirical results [1]. Since the models in this chapter have been influenced by several scientific models and since they are rather new, a good deal of future model alignment is in order. The neural network version can be compared with other neural models based on biological data and modified accordingly. Likewise, the mental process version can be compared with cognitive and behavioral models based on psychological data, and aligned accordingly. In addition, the operational rapid learning model could benefit from a careful statistical analysis of its learning efficiency and performance accuracy under changing conditions.

These rapid learning models may offer suggestions for scientific model development as well. They are represented by a *working* computing system that produces rapid learning and simple input-output performance. They are also linked to several scientific fields, thus providing connections between them. For example, speed and parallel processing concerns have influenced the statistical version of the rapid learning model. Such concerns might influence more general statistical theory as well. Also, statistical precision concerns have influenced the neural and mental process versions of the rapid learning model. Such concerns might suggest how and why biological and cognitive systems have evolved to their current form. Statistical analysis may offer other learning and cognitive explanations in rapid learning model terms as well. For example, the long-term influence of traumatic events on human personalities, even after extensive conditioning therapy, has long baffled clinical psychologists. This behavioral problem has an interesting explanation in terms of statistical "outliers," their influence on learning, and ways that their influence can be minimized.

Summary

Four concurrent learning and performance models are introduced in this chapter: (1) A rapid learning neuro-computing model is presented as a *working* system that learns from rapidly arriving data, while it provides monitoring, forecasting and control information. (2) A related neural network model is presented as an explanation for rapid learning during neuro-biological functioning. (3) A related cognitive processing model is presented as an explanation for simple mental activity in conjunction with stimulation, response, and paired associate learning. (4) A related statistical model is presented as an explanation for precise estimation in statistical terms, along with highly parallel learning and performance provisions in neuro-computing terms. Various interrelationships among these models — along with the scientific fields that have shaped them — appear to offer considerable food for future scientific thought.

REFERENCES

1. R.J. Jannarone, "Conjunctive Item Response Theory: Cognitive Research Prospects," in M. Wilson (Ed.), *Objective Measurement: Theory into Practice*, Vol.1, Ablex, Norwood, NJ, pp. 211-236, 1992.
2. R.J. Jannarone, "A Concurrent Learning and Performance Information Processing System," *RCNS Technical Report Series*, No. PAT94-01, Rapid Clip Neural Systems, Inc., Atlanta, GA, 1994.
3. R.J. Jannarone,"Concurrent Learning and Performance Analog Chip Design," *RCNS Technical Report Series*, No. PAT96-01, Rapid Clip Neural Systems, Inc., Atlanta, GA, 1996.
4. C. Mead, *Analog VLSI and Neural Systems*, Addison-Wesley, Reading, MA, 1995.
5. C. Mead and L. Conway, *Introduction to VLSI Systems*, Addison-Wesley, Reading, MA, 1980.
6. K. Naik, *A Concurrent Information Processing Parallel Design Analysis*, Unpublished Masters Thesis, University of South Carolina, 1996.
7. J.A. Deutsch & D. Deutch, *Physiology Psychology*, Dorsey, Homewood, IL, 1973.
8. D.E. Rumelhert & W.L. McClelland (Eds.), *Parallel Distributed Processing, Explorations in the Microstructure of Cognition,* Vol. 1, MIT Press, Cambridge, MA, 1986.
9. P. Wasserman, *Advanced Methods in Neural Computing*, Van Nostrand Reinhold, New York, 1993.
10. H. Gleitman, *Basic Psychology*, W.W. Norton, New York, 1983.
11. G. Tatman, R.J. Jannarone, & C.M. Amick, "Neural Networks for Speech Recognition: Contrasts Between a Traditional and a Parametric Approach," Unpublished Technical Report, Machine Cognition Laboratory, University of South Carolina, 1994.
12. R.D. Hawkins, T. Abrams, T.J. Carew, & E.R. Kandel, "A Cellular Mechanism of Classical Conditioning in Aplysia: Activity-Dependent Amplification of Presynaptic Facilitation," *Science*, Vol. 219, pp. 400-405, 1983.
13. S. Coren, C. Porac, & L.M. Ward, *Sensation and Perception*, Academic Press, New York, 1978.
14. J.M. Mendel, *Lessons in Estimation Theory for Signal Processing, Communications and Control*, Prentice-Hall PTR, Englewood Cliffs, NJ, 1995.
15. E.L. Lehmann, *Theory of Point Estimation*, Wiley, New York, 1983.

16. E.L. Lehmann, *Testing Statistical Hypotheses*, 2nd Edn., Wiley, New York, 1986.
17. P.J. Bickel & K.A. Doksum, *Mathematical Statistics: Basic Ideas and Selected Topics*, Holden-Day, San Francisco, 1977.
18. R.J. Jannarone,"The ABC of Measurement," Unpublished Technical Report, Machine Cognition Laboratory, University of South Carolina, 1994.
19. T.W. Anderson, *An Introduction to Multivariate Statistical Analysis*, 2nd Edn., Wiley, New York, 1984.
20. H. Scheffé, *The Analysis of Variance*, Wiley, New York, 1983.
21. D.R.. Cox, *The Analysis of Binary Data*, London, Methuen, 1966.
22. Y.M.M. Bishop, S.E. Fienberg, & P.W. Holland, *Discrete Multivariate Analysis: Theory and Practice*, MIT Press, Cambridge, MA, 1975.
23. L.A. Goodman, *Analyzing Qualitative/Categorical Data*, Jay Magidson, Cambridge, MA, 1978.
24. R.J. Jannarone, K.F. Yu, and Y. Takefuji, "Conjunctoids: Statistical Learning Modules for Binary Events," *Neural Networks*, Vol. 1, pp. 325-337, 1988.
25. R.J. Jannarone, "Conjunctive Item Response Theory Kernels," *Psychometrika*, Vol. 51, pp. 449-460, 1986.
26. R.J. Jannarone, "Locally dependent models: conjunctive item response theory," In W.J. van der Linden & R.K. Hambleton III (Eds.), *Handbook of Modern Item Response Theory*, Springer-Verlag, New York, pp. 465-480, 1996.
27. R.J. Jannarone, K.F. Yu, and J.E. Laughlin, "Easy Bayes Estimation for Rasch Type Models," *Psychometrika*, Vol. 55, pp. 449-460, 1990.

Part 3

Rapid Learning Foundations

Part 3 describes real-time learning functions that enable rapid learning system operation. The material in Part 3 starts with simple linear prediction, involving only one or two measurements per record, and proceeds to several varieties of multivariate prediction. Chapter 5 describes rapid learning from the perspective of its closest established counterpart — linear regression. Chapter 5 begins with a brief description of simple linear regression and ends with a brief description of multivariate linear regression. Chapter 6 describes the non-linear, spatial, and time series enhancements that give rapid learning methods a broad application domain. A technical background in statistics will ease understanding of the material in Part 3, although statistical expertise is not essential. Exceptions that require statistical expertise are clearly marked as being optional material for information processing specialists. Part 3 should give readers an understanding of rapid learning operations at a basic level, along with the ability to reproduce rapid learning results manually, given the time and inclination to do so.

5

Basic Concurrent Learning and Prediction

INTRODUCTION . . . 134
5.1. STATISTICAL ESTIMATION AND PREDICTION . . . 135
5.1.1. Univariate Estimation and Prediction . . . 135
5.1.2. Bivariate Estimation and Prediction . . . 137
5.1.3. Multivariate Estimation and Prediction . . . 141
5.1.4. Weighted Estimation and Prediction . . . 144
5.1.5. Basic Statistical Estimation Formulas . . . 146
5.2. SIMPLE CONCURRENT LEARNING AND PREDICTION . . . 149
5.2.1. Univariate Learning and Prediction . . . 149
5.2.2. Paired Associate Learning and Prediction . . . 154
5.2.3. Differential Reinforcement Learning . . . 159
5.3. MULTIVARIATE CONCURRENT LEARNING AND PREDICTION . . . 161
5.3.1. Solving a Certain Numerical Problem . . . 161
5.3.2. Learning and Prediction With Missing Measurements . . . 163
5.3.3. Basic Concurrent Learning and Prediction Formulas . . . 167
CONCLUSION . . . 170
REFERENCES . . . 171

INTRODUCTION

This chapter describes rapid learning and prediction when measurements are linearly related. The first section reviews sample-based linear statistical analysis, emphasizing aspects that are used by rapid learning methods. The next section describes rapid learning and prediction concepts for applications involving one or two variables. The final section describes rapid learning and prediction in more detail for applications involving several variables.

Most sections in this chapter are written at a conceptual level that describes input-output operation in general terms. A few sections are written at a mathematical level that describes operation in precise terms. The conceptual sections allow readers to understand how rapid learning methods perform linear prediction. The mathematical sections include formulas that readers may use to reproduce rapid learning results manually, if they have the inclination and time to do so. The mathematical sections, which are clearly labeled as such, contain only technical details that are not essential for applying rapid learning methods.

The results in this chapter are closely tied to standard results from statistical linear regression [1-2], along with results from Bayesian statistical theory [3-4]. Readers who are familiar with these areas of statistics will be better prepared than others to understand the chapter. However, readers without statistical backgrounds should be able to understand the essential material with moderate effort.

This chapter, like other chapters in Part 3 and Part 4, describes rapid learning in primarily functional terms. As a result, readers may wish to refer occasionally to related sections of chapters 1-4, which are more conceptual. References to appropriate earlier sections are provided throughout this chapter.

All sections in this chapter are based on data available in the form of records. Each record contains at least one measurement value from an instrument or a sensing device, and if a record contains more than one measurement the measurements are separated and ordered. The number and order of measurements are fixed from record to record. Each record contains one or more measurement values that have been sampled at the same time during a process or for the same individual in a population.

This sampling scheme applies to all previously presented examples: the strain gauge example involves regularly arriving records, each of which contains several strain gauge values (see section 1.1); the commodities index forecasting example involves regularly arriving records, each of which contains several commodity prices (see section 1.2); the missile control example involves records arriving every few milliseconds, each containing three position coordinates (see section 1.3); the pattern recognition problem involves visual pattern records, each containing 49 pixel values (see section 1.4) and the health monitoring example involves patient records, each of which contains measurements that may not be made at the same time, but are all made on the same patient. Consequently, the functional results from this chapter apply directly to all applications that are described elsewhere in this book.

This chapter should enable readers to describe how rapid learning methods perform concurrent learning and prediction in the linear case. Readers who wish to take the time do so should also be able to compute any linear prediction results manually that *Rapid Learner*™ software produces automatically. It should be noted, however, that while these formulas are equivalent mathematically to *Rapid Learner*™ formulas, they are far less efficient computationally. Readers are therefore advised to choose very simple problems for manual calculations because manually solving practical problems may require substantial effort.

5.1 STATISTICAL ESTIMATION AND PREDICTION

This section introduces sample-based, statistical estimation and prediction methods, which involve an estimation phase followed by a prediction phase. During the first phase, estimation records are gathered and analyzed in a sample. Estimation records have non-missing measurements, which are used to estimate prediction functions. During the second phase, prediction records are received, each of which has one or more measurements that must be predicted for monitoring, forecasting, or control purposes. This section describes linear prediction, for which predicted dependent variable values are additive combinations of independent variables. The section is essentially a review of standard results from linear regression that are needed to explain linear rapid learning methods, which are explained later in the chapter.

5.1.1 Univariate Estimation and Prediction

The univariate estimation case involves only one measurement in each estimation and prediction record. The method described in this section can be used to obtain a fixed estimate and tolerance band for monitoring measurements from a single sensing device. The method involves gathering an estimation sample, using the sample to estimate the mean and variance of the measurements, and using the estimates to compute a tolerance band. This method will provide accurate estimates and tolerance bands if measurement errors are normally distributed and they are stationary over time (see section 2.2).

Suppose that for the structural test monitoring example described in section 1.1, strain gauge measurements are available every second and test engineers have decided to use the first 15 records to compute sample estimates. For any strain gauge, the estimation sample can be represented as an array, which may be labeled as follows:

$$\boldsymbol{j}^{[\text{SAMPLE}]} = (j^{\{1\}}, j^{\{2\}}, \ldots, j^{\{15\}}).$$

A standard statistical approach to computing an estimate and a tolerance band based on this sample involves first computing the sample mean,

$$\hat{\mu} = (j^{\{1\}} + j^{\{2\}} + \cdots + j^{\{15\}})/15,$$

along with the sample error variance,

$$\hat{\nu}^{[E]} = [(j^{\{1\}} - \hat{\mu})^2 + (j^{\{2\}} - \hat{\mu})^2 + \cdots + (j^{\{15\}} - \hat{\mu})^2]/15.$$

An estimate and tolerance band for future monitoring purposes are then computed as follows. For future time points t' after the estimation sample time points ($t' = 16, 17, \ldots$), prediction estimates have the form,

$$\hat{j}^{\{t'\}} = \hat{\mu}$$

and tolerance bands have the form,

$$(\underline{j}^{\{t'\}}, \bar{j}^{\{t'\}}) = (\hat{j}^{\{t'\}} - c\hat{\nu}^{[E]1/2}, \hat{j}^{\{t'\}} + c\hat{\nu}^{[E]1/2}).$$

Depending on the application, the constant c is chosen for suitably sensitive monitoring purposes. Most commonly $c = 2$ is chosen, resulting in a 95% tolerance band with the following "confidence interval" interpretation [5-6]: if (normal error and stationary sampling) assumptions are satisfied, then 95% tolerance bands may be expected to contain 95% of all future measurement values. Consequently, if a measurement falls outside its 95% tolerance band, the measurement is considered likely to be deviant.

Once each strain gauge estimate and tolerance band are obtained in this way, they may be used for monitoring future measurements for that gauge. In statistical sampling terms, all such future measurements are treated as prediction records. For each prediction record, the strain gauge value is obtained and compared to its predicted value and its tolerance band. If the measurement falls outside its tolerance band, it is considered deviant and appropriate corrective action is initiated.

As an example, Figure 5.1.1 contains the first 15 measurements that might be obtained from a strain gauge during a structural test (see section 1.1). The resulting estimates and 95% confidence intervals from using the above formulas are

$$\hat{j}^{\{t'\}} = 3.33,$$

and

Trial Number	Observed Value
1	3.2
2	0.1
3	3.8
4	2.2
5	0.7
6	1.7
7	4.1
8	5.3
9	2.8
10	6.1
11	5.4
12	2.0
13	3.7
14	5.8
15	3.1

FIGURE 5.1.1 A Univariate Estimation Sample
(Courtesy of Rapid Clip Neural Systems, Inc.)

$$(\underline{j}^{\{t'\}}, \bar{j}^{\{t'\}}) = (-0.07, 6.67)$$

respectively $(t' = 16, 17, \ldots)$. These estimates are relatively imprecise when compared to multivariate estimation and rapid learning alternatives, as is demonstrated later in this chapter.

5.1.2 Bivariate Estimation and Prediction

In most applications, univariate estimation and prediction may be improved by using alternative methods, which predict measurements as functions of other measurements. This sub-section considers the simplest such alternative, based on predicting a measurement value as a function of only one other measurement value. This is called the bivariate case, because it involves two variables per record.

Extending the strain gauge monitoring example from the previous section to the bivariate case, suppose that measurements for two strain gauges are obtained at each of the 15 time points. The estimation sample may then be represented by the following array:

$$j^{[SAMPLE]} = \begin{bmatrix} j^{[I]\{1\}} & j^{[D]\{1\}} \\ j^{[I]\{2\}} & j^{[D]\{2\}} \\ \vdots & \vdots \\ j^{[I]\{15\}} & j^{[D]\{15\}} \end{bmatrix}.$$

As in the previous section, the superscript numbers in braces are time point labels. Each row in this array thus corresponds to an estimation record made up of two measurements. The first measurement in each record is used to predict the second measurement, in order to improve prediction precision over univariate estimation. The first column is thus made up of independent variable measurements labeled by [I] superscripts, and the second column is made up of dependent variable measurements labeled by [D] superscripts.

A standard statistical approach to computing an estimation function and tolerance bands based on this sample involves first computing the following statistics:

1. The sample mean vector, $\hat{\mu} = (\hat{\mu}^{[I]}, \hat{\mu}^{[D]})$, where each of the two means in the vector is computed just as the mean in the previous section.
2. The sample variance vector, $\hat{\nu} = (\hat{\nu}^{[I]}, \hat{\nu}^{[D]})$, where each of the two variances in the vector is computed just as the variances in the previous section.
3. The sample covariance value,

$$\hat{\nu}^{[ID]} = [(j^{[I]\{1\}} - \hat{\mu}^{[I]})(j^{[D]\{1\}} - \hat{\mu}^{[D]}) +$$

$$(j^{[I]\{2\}} - \hat{\mu}^{[I]})(j^{[D]\{2\}} - \hat{\mu}^{[D]}) + \cdots +$$

$$(j^{[I]\{15\}} - \hat{\mu}^{[I]})(j^{[D]\{15\}} - \hat{\mu}^{[D]})]/15.$$

Once these statistics are computed, a prediction function is created of the form,

$$\hat{j}^{[D]\{t'\}} = \hat{\mu}^{[D]} + (j^{[I]\{t'\}} - \hat{\mu}^{[I]})\hat{\rho}^{[ID]},$$

where the sample regression weight $\hat{\rho}^{[ID]}$ is computed as follows:

$$\hat{\rho}^{[ID]} = \hat{\nu}^{[ID]}/\hat{\nu}^{[I]}.$$

In addition, tolerance band functions are computed as follows:

$$(\underline{j}^{[D]\{t'\}}, \bar{j}^{[D]\{t'\}}) = (\hat{j}^{[D]\{t'\}} - c\hat{\nu}^{[DE]1/2}, \hat{j}^{[D]\{t'\}} + c\hat{\nu}^{[DE]1/2}),$$

where the dependent variable error variance $\hat{\nu}^{[DE]}$ is computed as follows:

$$\hat{\nu}^{[DE]} = \hat{\nu}^{[D]} - \hat{\nu}^{[ID]2}/\hat{\nu}^{[I]}.$$

These bivariate estimation and tolerance band functions differ from their univariate counterparts in one important way. The univariate estimate and tolerance band are fixed, but the bivariate versions vary linearly with values of independent variable measurements. By being variable in this way, the bivariate estimates are more precise, to the extent that independent and dependent variables are linearly correlated. In many applications such as strain gauge monitoring this increased precision can be substantial as the next example shows.

As an extension of the univariate example from the previous section, Figure 5.1.2.1 contains the first 15 pairs of measurements that might be obtained from two strain gauges during a structural test. The independent variable measurements are the same as the 15 measurements from the previous example. Hence, the following bivariate results are directly comparable to the previous univariate results. Based on this sample, the computed 95% tolerance band is

Trial Number	Observed Values 1	2
1	9.7	3.2
2	7.3	0.1
3	11.0	3.8
4	8.7	2.2
5	8.3	0.7
6	6.8	1.7
7	10.8	4.1
8	13.3	5.3
9	9.2	2.8
10	14.9	6.1
11	12.6	5.4
12	7.4	2.0
13	11.5	3.7
14	14.6	5.8
15	8.9	3.1

FIGURE 5.1.2.1 A Bivariate Estimation Sample
(Courtesy of Rapid Clip Neural Systems, Inc.)

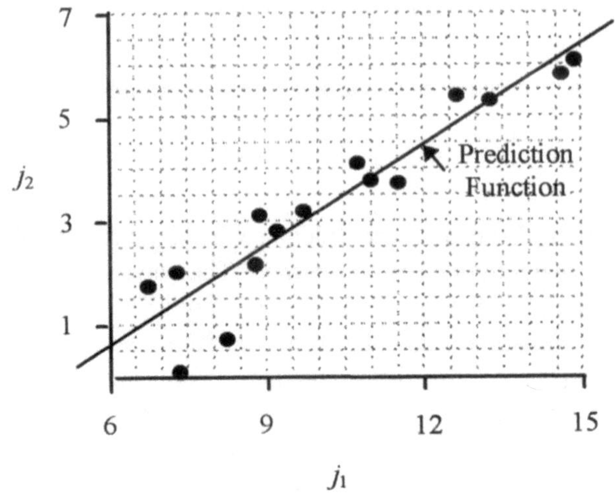

FIGURE 5.1.2.2 Bivariate Statistical Estimation
(Courtesy of Rapid Clip Neural Systems, Inc.)

$$(\underline{j}^{[D]\{t'\}}, \bar{j}^{[D]\{t'\}}) = (j^{[I]\{t'\}} - 0.60,\ j^{[I]\{t'\}} + 0.60).$$

In comparison with the univariate 95% tolerance band from the previous section, this tolerance band is narrower by a factor of 0.18. In practical terms, this means that bivariate monitoring may be considerably more valuable. For example, suppose that a gauge is beginning to break down to the extent that its bivariate tolerance band indicates deviant values but its univariate values do not. The resulting early warning that the bivariate tolerance band can provide but the univariate band cannot provide may produce substantial savings in terms of preventing costly accidents and down time.

Figure 5.1.2.2 shows bivariate statistical estimation in graphical terms, based on the Figure 5.1.2.1 data. The vertical and horizontal axes represent dependent gauge measurement values, respectively. The points represent the 15 pairs of sample measurements in Figure 5.1.2.1, and the *regression line* represents the prediction function that was determined from the sample.

Figure 5.1.2.3 shows bivariate statistical prediction in graphical terms, along the same lines as Figure 5.1.2.2. Along with the same regression line, the figure represents the bivariate tolerance band by short dashed lines, as well as the univariate tolerance band by long dashed lines. The figure also shows two bivariate prediction points. The point shown in the lower left lies within the bivariate tolerance band, indicating that the observed dependent gauge measurement value was not deviant for that point. The point in the upper right lies outside the bivar-

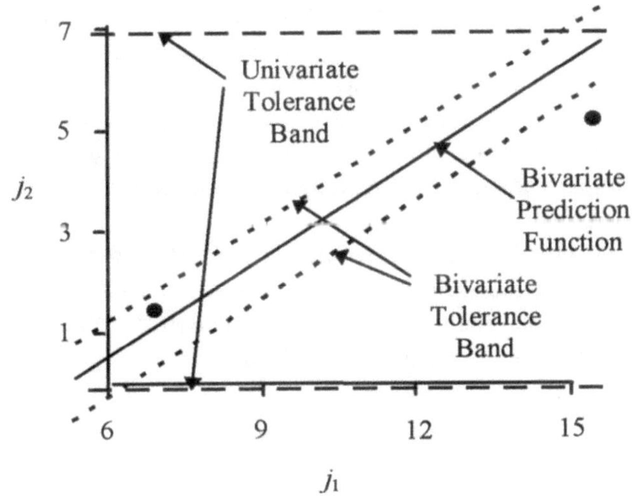

FIGURE 5.1.2.3 Bivariate Statistical Prediction
(Courtesy of Rapid Clip Neural Systems, Inc.)

iate tolerance band but inside the univariate tolerance band. In keeping with the above discussion, the point in the upper right represents a case where bivariate prediction may add precision to univariate prediction, as well as adding sensitivity to univariate monitoring.

5.1.3 Multivariate Estimation and Prediction

Just as bivariate methods may improve precision over univariate methods, multivariate methods using several dependent variables may improve prediction even more. This sub-section describes sample-based multivariate prediction in terms of features, benefits, and practical concerns. (A detailed formulation of multivariate regression is provided in section 5.1.5.) The description first outlines the use of several independent variables to predict one dependent variable and then outlines the prediction of several dependent variables in each record.

Extending the structural test monitoring example further, suppose that each record is represented as a vector containing 250 measurements (see section 4.1). Suppose further that each record at each time point t' is labeled as follows ($t' = 1, 2, \ldots$):

$$\mathbf{j}^{\{t'\}} = (j_1^{\{t'\}}, j_2^{\{t'\}}, \ldots, j_{250}^{\{t'\}}).$$

In terms of sample-based regression, then, the practical strain gauge monitoring problem that is outlined in section 1.1 may be represented as follows: First, a representative sample of measurements obtained. Next, the same formulas that appear earlier in this section are used to compute required sample statistics for computing linear prediction functions. These statistics include 250 sample means, 250 sample variances, and a sample covariance for each pair of gauges (31,125 covariances in all).

Next, these statistics are used to predict one set of regression weights for predicting each gauge measurement as a linear function of all other gauge measurements. For example, the linear regression function for predicting gauge 250 as a function of the other 249 gauges has the form,

$$\hat{j}_{250}^{\{t'\}} = \hat{\mu}_{250} + (j_1^{\{t'\}} - \hat{\mu}_1)\hat{\rho}_{1,250}$$

$$+ (j_2^{\{t'\}} - \hat{\mu}_2)\hat{\rho}_{2,250}$$

$$+ \cdots + (j_{249}^{\{t'\}} - \hat{\mu}_{249})\hat{\rho}_{249,250}$$

($t' = 251, 252, \ldots$). Obtaining this equation, like the other 249 prediction equations, requires computing 249 regression weights from sample statistics. A tolerance band is computed for each prediction function as well, based on sample statistics.

Once a sample has been gathered and these prediction functions have been computed from sample statistics, they may be used for monitoring, just as in the simpler cases presented earlier in this section. Each time a strain gauge measurement record arrives, each predicted value may be computed and compared with its measured value. If the measured value falls outside its tolerance band, it is recognized as deviant in keeping with earlier discussion.

Just as bivariate prediction may produce narrower tolerance bands than univariate prediction, prediction with such large numbers of independent variables may produce even narrower tolerance bands. However, before precise monitoring benefits can be achieved from using many independent variables, some potentially difficult problems must be solved. First, standard regression procedures require sample sizes containing at least as many records as the number of measurements that each record contains. This requirement is not a problem in some applications with large prior samples available, but it is usually a problem when rapid learning is required. It is also a problem when many measurements per record are utilized as in questionnaire-based decision making.

Second, standard regression computations produce unstable results when many highly correlated variables are used. This problem may be overcome by employing step-wise regression procedures to remove redundant variables [1-2], but these procedures require expensive and time consuming expert effort. Third, even when such problems are absent, many hours of effort are often needed to create many prediction functions using standard regression tools. Finally, the biggest potential problem of all for rapid learning applications is the need to gather a training sample.

The necessary statistics background for understanding rapid learning linear prediction has now been presented, with two exceptions: the use of weighted estimation procedures in regression, which is presented in the next section; and problems associated with predicting several dependent variables at the same time, which will be outlined next. Since the formulas associated with the following description are detailed, they are presented in a separate optional section (see section 5.1.5).

In many applications including most that are presented in chapter 1, more than one variable must be predicted at the same time. Standard multiple regression procedures can be used to predict each measurement as a function of all others, if all measurements are available at the same time. Such is the case in many monitoring cases such as all those described in chapter 1 (see section 1.1). In many forecasting and control cases, however, several variables must be predicted at once, each of which has not yet been measured. In such cases, standard multivariate regression procedures may be used. Multivariate regression prediction functions resemble multiple regression functions, in that each prediction function is an additive combination of independent variables. Likewise, multivariate regression uses regression weights that must be estimated from a sample. Multiple and multivariate regression weights may differ substantially, however, especially if the dependent variables are highly correlated. As a result, prediction functions from multiple regression may not be accurate in multivariate prediction settings.

Inaccuracies associated with multiple regression prediction in multivariate regression settings pose several problems in prediction applications. Prediction problems often occur in rapid response settings where independent variables are missing in estimation records and/or prediction records. For example, suppose that multiple regression estimates from a sample are used to predict an aircraft trajectory as a function of three radar signals, but one of the radar signals breaks down during tracking. In the interest of speed, it may be tempting to predict missile trajectory using multiple regression weights that have been previously estimated, with use restricted to terms involving only non-missing radar signals (this is easy to do in practice by replacing missing signals with their mean). However, highly distorted results can be expected if this approach is taken in this particular application. Similar problems hold in settings where estimation samples must be analyzed quickly and occasional missing values may be present (see section 5.3.3).

A brute-force way to deal with missing value problems is to anticipate all possible missing variable patterns and to predict regression weights separately for all such patterns. This approach is only practical when a few missing data patterns prevail. As a better alternative, adjustments may be made to multiple regression weights that have been estimated from non-missing data. Such adjustments may be tailored to suit any independent variable missing value pattern from trial to trail, using standard multivariate regression formulas (see section 5.1.5). However, these formulas require matrix inversion, which may cause costly time delays in rapid prediction settings such as aircraft tracking.

Rapid learning methods include provisions for dealing with missing values by adjusting prediction functions very quickly (see section 5.3). Rapid learning adjustments produce the same mathematical results as multivariate regression adjustments (see section 5.1.5), but much more quickly. Unlike standard statistical adjustments, *Rapid Learner*™ adjustment algorithms do not require matrix inversion, and they are designed for special-purpose parallel processors (resembling the parallel Kernel processor — see section 4.2). Along with added missing-value handling capacity, these algorithms have uses in refinement operations such as adjusting connection weights after removing redundant variables. As with other rapid learning features, then, this provision allows necessary adjustments to be made *during* rapidly changing conditions. In this case, the resulting benefit is being able to adjust for missing measurements or deleted variables on the fly (see section 5.3.3).

5.1.4 Weighted Estimation and Prediction

Regression methods have standard provisions for <u>weighting</u> estimation records differentially [1]. These methods are reviewed in detail here, because rapid learning methods include equivalent provisions for differential learning. In both cases, weights are imposed on records in such a way that records with higher weights have more impact on estimation or prediction than records with lower weights have.

Continuing with the strain gauge monitoring example, suppose that estimation records for a structural test are obtained every second, as described earlier (see section 5.1.1). In that case, the most recently observed estimation records may best reflect relationships among measurements in the prediction sample, especially if relationships are changing gradually. For example, if load values are increasing throughout a test and measurements from the first 15 seconds are used for estimation, then means values that are weighted more toward recent estimation records will be better than mean values that are equally weighted.

The practical consequences of weighted statistical estimation are straightforward. Suppose that an estimation record has a weight value of 10. Then the effect of that record on mean estimates and other regression estimates will be the

Trial Number	Observed Values 1	2	Impact Values
1	9.7	3.2	1.0
2	7.3	0.1	1.0
3	11.0	3.8	1.0
4	8.7	2.2	1.0
5	8.3	0.7	1.0
6	6.8	1.7	2.0
7	10.8	4.1	2.0
8	13.3	5.3	2.0
9	9.2	2.8	2.0
10	14.9	6.1	2.0
11	12.6	5.4	4.0
12	7.4	2.0	4.0
13	11.5	3.7	4.0
14	14.6	5.8	4.0
15	8.9	3.1	4.0

FIGURE 5.1.4.1 Weighted Estimation Example Data
(Courtesy of Rapid Clip Neural Systems, Inc.)

same as if 10 records were in the estimation sample in place of that record. The same interpretation holds for other weight values as well. In general, an estimation record weight w has the same effect as w estimation records, each having a weight value of 1. In this way, the weight value w for an estimation record can be interpreted as a relative sample size for the record. Also, the ratio of w values for two different estimation records can be interpreted as the corresponding sample size ratio for the two records.

Figure 5.1.4.1 gives an example of one weighting scheme for the records that were presented earlier in Figure 5.1.2.2. The last 5 observation weights are twice the middle 5 weights, which in turn are twice the first 5 weights. According to the above interpretation, resulting weighted estimates are the same as if 35 estimation records had been used: the first 5 records, along with 2 sets of the next 5 records, along with 4 sets of the last 5 records.

Figure 5.1.4.2 shows the prediction function that results from using the weighted estimation samples in Figure 5.1.4.1. The function is noticeably different, compared with its unweighted counterpart in Figure 5.1.2.3. Most notably, the weighted function produces higher predicted values than the non-weighted version throughout the plot. Higher predicted values result because the dependent variable mean estimate $\hat{\mu}^{[D]}$ is higher in the weighted case, reflecting the higher values of more recent, more highly weighted measurements.

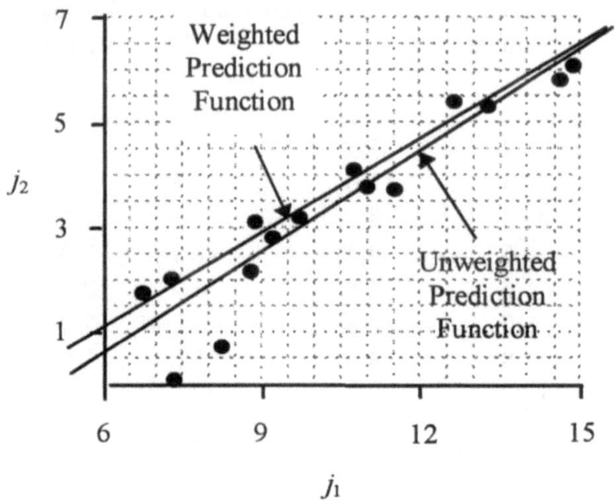

FIGURE 5.1.4.2 Weighted Statistical Prediction
(Courtesy of Rapid Clip Neural Systems, Inc.)

For this example, assigning larger weights to more recent sample measurements makes sense, because load levels are increasing throughout the test and the first several observations are used for estimation. Indeed, assigning higher weights to more recent measurements is sensible in most non-stationary settings (see section 4.4). Weighting makes less sense in stationary settings because estimation records are not expected to differ substantively from each other. Weighting *is* an important component of rapid learning systems, however, because they are designed for use in non-stationary settings. For this reason, effort has been devoted to producing *Rapid Learner*™ weighted learning options, with the same simple interpretation as sample-based weighting options (see sections 3.1.7 and 5.2.3).

5.1.5 Basic Statistical Estimation Formulas

(This sub-section covers mathematical details that are not essential for practical use of rapid learning methods.)

The following notation conventions are followed in this and other formulation sections throughout this book:

- Greek letters are parameters to be estimated.
- Roman letters are observable measurements.

- Caret overscripts (e.g., $\hat{v}$ denote sample-based, (weighted) least-squares parameter estimates [1].
- cap overscripts (e.g., $\widehat{j}^{\{t'\}}$) denote rapid learning measurement prediction estimates.
- Bold face letters are arrays.
- Entries in subscript parentheses are array dimensions [7].
- One-dimensional arrays are row vectors.
- Entries in superscript braces (e.g., $v^{\{t'\}}$) are time points or record labels.
- Entries in superscript brackets (e.g., $v^{[\mathrm{I}]}$) are labels.
- **T** superscripts (e.g., $\boldsymbol{e}^{\mathrm{T}}$) denote array transposition [7].
- **1** denotes an array of 1's.
- **I** denotes the identity matrix [7].
- $\widetilde{\mathbf{D}}(\boldsymbol{x})$ denotes a diagonal matrix containing the elements of the vector $\boldsymbol{x}$ on its main diagonal [7].
- Prime superscripts (e.g., t') are counters (dummy variables).
- Plus superscripts(e.g., j^+) are maximum values.

Consider the case where estimation and prediction records each contain i^+ independent variable measurements and d^+ dependent variable measurements. The general multivariate regression case is covered in this section where $i^+ > 0$ and $d^+ > 0$. The multiple regression special case corresponds to $d^+ = 1$ with $i^+ > 1$ (see section 5.1.4), the bivariate case corresponds to $i^+ = d^+ = 1$ (see section 5.1.2), and the univariate case corresponds to $d^+ = 1$ with $i^+ = 0$ (see section 5.1.1).

Suppose that t^+ estimation records are available that are denoted by

$$\boldsymbol{j}_{(j^+)}^{\{t'\}} = (\boldsymbol{j}_{(i^+)}^{[\mathrm{I}]\{t'\}}, \boldsymbol{j}_{(d^+)}^{[\mathrm{D}]\{t'\}}),\ t' = 1, \ldots, t^+,$$

and assembled into an estimation sample array,

$$\boldsymbol{j}_{(t^+ \times j^+)} = \begin{bmatrix} \boldsymbol{j}^{\{1\}} \\ \vdots \\ \boldsymbol{j}^{\{t^+\}} \end{bmatrix}.$$

Suppose further that prediction records are provided for every time point after t^+, having non-missing $\boldsymbol{j}^{[\mathrm{I}]\{t'\}}$ values ($t' = t^+ + 1, \ldots$). Standard (weighted) multivariate regression prediction estimates for each record then have the form,

$$\hat{j}^{[\mathrm{D}]\{t'\}}_{(d^+)} = \hat{\mu}^{[\mathrm{D}]}_{(d^+)} + (j^{[\mathrm{I}]\{t'\}}_{(i^+)} - \hat{\mu}^{[\mathrm{I}]}_{(i^+)})\hat{\rho}^{[\mathrm{ID}]}_{(i^+ \times d^+)},$$

where

$$\hat{\mu}_{(j^+)} = (\hat{\mu}^{[\mathrm{I}]}_{(i^+)}, \hat{\mu}^{[\mathrm{D}]}_{(d^+)})$$

$$= (\boldsymbol{w}_{(t^+)}\mathbf{1}_{(t^+ \times 1)})^{-1}\boldsymbol{w}_{(t^+)}\boldsymbol{j}_{(t^+ \times j^+)},$$

$$\hat{\rho}^{[\mathrm{ID}]}_{(i^+ \times d^+)} = \hat{v}^{[\mathrm{II}]-1}_{(i^+ \times i^+)}\hat{v}^{[\mathrm{ID}]}_{(i^+ \times d^+)},$$

and

$$\hat{v}_{(j^+ \times j^+)} = (\boldsymbol{w}_{(t^+)}\mathbf{1}_{(t^+ \times 1)})^{-1}\left[\boldsymbol{j}_{(t^+ \times j^+)} - \mathbf{1}_{(t^+ \times 1)}\hat{\mu}_{(j^+)}\right]^{\mathrm{T}} \widetilde{\mathbf{D}}(\boldsymbol{w})^{-1}\left[\boldsymbol{j}_{(t^+ \times j^+)} - \mathbf{1}_{(t^+ \times 1)}\hat{\mu}_{(j^+)}\right]$$

$$= \begin{bmatrix} \hat{v}^{[\mathrm{II}]}_{(i^+ \times i^+)} & \hat{v}^{[\mathrm{ID}]}_{(i^+ \times d^+)} \\ \hat{v}^{[\mathrm{YX}]}_{(d^+ \times i^+)} & \hat{v}^{[\mathrm{DD}]}_{(d^+ \times d^+)} \end{bmatrix}.$$

The above formulation is routinely implemented by multivariate regression software in standard statistical packages such as SAS™ [8]. As shown later in this chapter (see section 5.3.5), basic concurrent learning functions may be implemented in precisely the same terms. Consequently, standard statistical packages may be used to produce precisely the same results as *Rapid Learner*™ software results. However, the time required to obtain standard results may be considerable. For example, strain gauge monitoring estimation could be accomplished by SAS™ in the following way. At each time point t' ($t' = t^+ + 1, \ldots$), a new estimation record $\boldsymbol{j}^{\{t'\}}$ may be observed having no missing measurements. The estimation sample array $\boldsymbol{j}$ may then be augmented to include $\boldsymbol{j}^{\{t'\}}$ along with an appropriate $w_{t'}$ weight value, after which t^+ may be incremented by 1 and prediction function estimates may be recomputed. In this monitoring case, one prediction function would be computed separately for each strain gauge. Once the prediction functions were estimated in this way, they would be used to predict each strain gauge value in order to compute monitoring statistics at the next time point. Creating each such function at each time point by brute force requires fitting a new model to the estimation sample data for each measurement at each time point.

The term "brute force" used in this context is of course relative. Performing the above operations using statistical packages requires substantial expertise and produces results orders of magnitudes faster than hand calculations would require. Furthermore, many of these operations can be automated and streamlined to a large degree. When compared to rapid learning software, however, the term makes more sense. *Rapid Learner*™ software can perform all of the above operations automatically *in less than one second per record*, when $j^{+} = 1{,}000$ (see section 2.4.1).

5.2 SIMPLE CONCURRENT LEARNING AND PREDICTION

This section introduces basic concurrent learning for the relatively simple cases where only one or two variables are measured per record. Just as the statistical estimation description in the previous section, this section describes settings where one record arrives at every time point. However, this section describes learning and prediction *both* occurring at each time point. Also, this section deals with prediction functions that are updated recursively from trial to trial, instead of being computed once from an estimation sample. The results in this section are explained in terms of the statistical results from the previous section, so that readers can interpret concurrent information processing operations in precise statistical terms. In addition, the section outlines how the rapid learning system uses recursive updating to learn so quickly.

5.2.1 Univariate Learning and Prediction

This sub-section follows the univariate statistical estimation description given earlier (see section 5.1.1), as well as the weighted estimation description given earlier (see section 5.1.4). The measurement values for the example below are the same as those for the earlier univariate example, and the weights are the same as those for the weighted estimation example. Notation in this sub-section follows notation in these two earlier descriptions as well. Consequently, both earlier descriptions might best be reviewed before reading this sub-section.

The following description begins with an overview of key rapid learning concepts, followed by an example giving rapid learning results, followed by a summary of rapid learning formulas for the univariate case. The description ends with details regarding initialization, including a *prior sample* explanation of initial values that is useful for interpreting rapid learning results. To keep the description manageable, the following results are based on measurements all having plausibility values of 1 (see section 4.2 — the more general case is covered in section 5.3)

The rapid learning approach to predicting in the univariate case has similarities and differences with the sample-based statistical approach (see section 5.1.1).

First, at each time point t' ($t' = 1, 2, \ldots$), two rapid learning parameter estimates are available for predicting the current measurement value as well as its tolerance band. These two estimates, which are denoted by $\mu^{\{t'-1\}}$ and $\nu^{[E]\{t'-1\}}$, correspond to the sample mean $\hat{\mu}$ and the sample error variance $\hat{\nu}^{[E]}$. Rapid learning parameter estimates differ, however, in that they are updated recursively at each time point, rather than being computed once from a sample (see the end of this sub-section).

Second, rapid learning predicted estimates, which are denoted by $\hat{j}^{\{t'\}}$, are computed based on prior parameter estimates just as their sample-based counterparts $\hat{j}^{\{t'\}}$. Unlike sample-based estimates, however, they are computed at each time point from the first point onward, instead of being computed only after sample-based parameter estimation. Each rapid learning estimate at each time point is based on parameter estimates that have been computed up to, but not including, that time point. As a result, initial parameter values must be provided prior to the first trial and they must be updated at the end of each trial.

Third, initial parameter values may either be explicitly provided by *Rapid Learner*™ software users, or the software may supply them by default (see section 3.1). If the default approach is selected, one record is read prior to rapid learning operation, and the measurement mean is set to its first measurement value. Selecting the default also produces an initial error variance value of 1. This initializing approach avoids producing parameter estimates during early learning that may be strongly influenced by arbitrary initial values (see the end of this sub-section for an example).

Fourth, the sample-based error variance estimate $\hat{\nu}^{[E]}$ is an average among squared differences between sample observations and the sample mean (see section 5.1.1). In contrast, the rapid learning error variance estimate $\nu^{[E]\{t'-1\}}$ is an average squared difference between each measurement up to the current time point and its predicted estimate at that time point (see the end of this sub-section). As a result, $\nu^{[E]\{t'-1\}}$ reflects prediction accuracy for all previous records rather than prediction accuracy only for sample records.

Finally, sample-based impact weights $w_{t'}$ are assigned explicitly to every prediction record (see section 5.1.4). In contrast, rapid learning impact weights $w^{\{t'\}}$ are adjusted implicitly and recursively from the first time point onward. At each time point t' ($t' = 1, 2, \ldots$), a learning weight $l^{\{t'\}}$ establishes the impact of the current record, relative to the sum of all prior record impacts (see the end of this sub-section). If $l^{\{t'\}}$ is large, all prior impact weights become reduced more than if $l^{\{t'\}}$ is small. This use of recursive learning weights allows a variety of learning schedules to be imposed for a variety of applications (see section 5.2.3). Recursive learning weights also provide the flexibility to refine models quickly when measurement correlations change (see chapter 7).

Trial Number	Impact Weight	Learning Weight	Observed Value	Predicted Value	Tolerance Band
0	1.0	—	3.2	—	—
1	1.0	1	0.1	3.20	(1.20, 5.20)
2	1.0	1/2	3.8	1.65	(–2.96, 6.25)
3	1.0	1/3	2.2	2.36	(–2.14, 6.87)
4	1.0	1/4	0.7	2.33	(–1.58, 6.23)
5	2.0	2/5	1.7	2.00	(–1.78, 5.78)
6	2.0	2/7	4.1	1.95	(–1.51, 5.41)
7	2.0	2/9	5.3	2.56	(–0.15, 6.29)
8	2.0	2/11	2.8	3.17	(–1.00, 7.35)
9	2.0	2/13	6.1	3.10	(–0.68, 6.89)
10	4.0	4/15	5.4	3.57	(–0.64, 7.77)
11	4.0	4/19	2.0	3.81	(–0.33, 7.95)
12	4.0	4/23	3.7	3.43	(–0.60, 7.46)
13	4.0	4/27	5.8	3.48	(–0.20, 7.15)
14	4.0	4/31	3.1	3.82	(–0.01, 7.65)

FIGURE 5.2.1 A Univariate Rapid Learning Example
(Courtesy of Rapid Clip Neural Systems, Inc.)

Figure 5.2.1 gives observed and predicted measurements for one strain gauge in the univariate case. Each row in the figure corresponds to one time point from 0 to 15. From time point 1 onward, a measurement record is received containing one measurement value (measured values from points 16 onward are not shown). At each time point, a weight is assigned to each value as well (the weight for point 15 is not shown). In addition, at each time point from 1 onward, a predicted value is computed, which is not shown. However, the tolerance band that is centered around each predicted value is shown.

The impact weights assigned to trials 0 through 14 in Figure 5.2.1 are the same as the weights assigned to trials 1-15 for the previous weighted estimation example (see section 5.1.3). The measurement values for trials 1 through 14 are the same as measurement values 2-15 for the previous univariate estimation example (see section 5.1.1).

The results in Figure 5.2.1 are based on recursively updating parameter values from time point 1 onward, along with initializing parameter values at time point 0. The initial value of the mean parameter $\mu^{\{0\}}$ is set to the value of the first observation, which is the same as measurement value 1 for the previous univariate estimation example (see section 5.1.1). The initial value of the error variance parameter $\nu^{[E]\{0\}}$ is set to 1. The tolerance band values of (–2.0, 2.0) at time point 0 reflect that initial value along with a tolerance band factor of 2, corresponding to a 95% confidence interval (see section 5.1.1).

When compared with corresponding sample-based results (see section 5.1.1), the rapid learning results of Figure 5.2.1 show some interesting contrasts, including the following:

- Predicted values are available from point 1 onward.
- Predicted values shift toward recently observed values from point 1 onward.
- Tolerance band values are available from point 1 onward.
- Tolerance bands shift as the process changes.
- Tolerance bands become narrower as the process continues.

Rapid learning formulas for the univariate case are given next. (Verifying by hand that these formulas produce columns 3 and 5 of Figure 5.2.1 is recommended as an optional exercise.) The rapid learning mean and variance are updated at each time point t' ($t' = 1, 2, \ldots$), as follows:

$$\mu^{\{t'\}} = (l^{\{t'\}} j^{\{t'\}} + \mu^{\{t'-1\}})/(1 + l^{\{t'\}})$$

and

$$\nu^{[\mathrm{E}]\{t'\}} = [l^{\{t'\}}(j^{\{t'\}} - \hat{j}^{\{t'\}})^2 + \nu^{[\mathrm{E}]\{t'-1\}}]/(1 + l^{\{t'\}}),$$

respectively, where

$$l^{\{t'\}} = w^{\{t'\}}/(w^{\{0\}} + w^{\{1\}} + \cdots + w^{\{t'-1\}})$$

is the current learning weight,

$$\hat{j}^{\{t'\}} = \mu^{\{t'-1\}}$$

is the current predicted value, $\mu^{\{0\}} = 3.2$ is the initial mean, and $\nu^{[\mathrm{E}]\{0\}} = 1$ is the initial variance.

Setting the initial mean to the first measurement value and the initial variance to 1 avoids extreme tolerance band values during early concurrent operation. For example, the Figure 5.2.1 trial 2 tolerance band was computed as follows:

$$\begin{aligned}(\underline{j}^{\{2\}}, \bar{j}^{\{2\}}) &= (\hat{j}^{\{2\}} - 2\nu^{[\mathrm{E}]\{1\}1/2}, \hat{j}^{\{2\}} + 2\nu^{[\mathrm{E}]\{1\}1/2}) \\ &= (1.65 - 4.6, \hat{j}^{\{t'\}} + 4.6),\end{aligned}$$

based on

$$\begin{aligned} v^{[E]\{1\}} &= [l^{\{1\}}(j^{\{1\}} - \hat{j}^{\{1\}})^2 + v^{[E]\{0\}}]/(1 + l^{\{1\}}) \\ &= [(0.1 - 3.2)^2 + 1.0]/2 \\ &= 5.30 \end{aligned}$$

and

$$\begin{aligned} \hat{j}^{\{1\}} &= \mu^{\{0\}} \\ &= 3.2. \end{aligned}$$

If an initial mean of 0 had been used instead, the tolerance band could have been much higher. For example, if room temperature readings were being monitored and if the first observed temperature was 72 degrees, then the first variance estimate would be

$$\begin{aligned} v^{[E]\{1\}} &= [(72 - 0)^2 + 1]/2 \\ &= 2{,}592.5, \end{aligned}$$

a high value due strictly to the arbitrary location of the temperature scale. The effect, then, of using the first observation to set the prior mean is to establish from the onset variance values based on squared differences between actual measurements, rather than differences between two numbers that may be on a different scale. The same initializing approach is used in the multivariate case in order to avoid similar scaling problems.

Rapid learning estimation may be interpreted as combining information from two sources at every time point: information from a *prior sample*, along with information from the current record. (This explanation is equivalent to standard interpretations of estimates based on Bayesian statistical inference [3-4].) For example, $\mu^{\{1\}}$ is equivalent to a weighted average of $j^{\{1\}}$ and the mean $\mu^{\{0\}}$ of measurements from an initial sample. Likewise, $v^{[E]\{1\}}$ is equivalent of a weighted average of $(\hat{j}^{\{1\}} - j^{\{1\}})^2$ and average squared differences between observed and predicted values $v^{[E]\{0\}}$ from an initial sample. In particular, the $\mu^{\{0\}}$ value of 3.2 and the $v^{[E]\{0\}}$ value of 1.0 in Figure 5.2.1 are unweighted sample mean and error variance values from two estimation records, having values 3.2−1 and 3.2+1. The same kind of interpretation can be given to results at every time point. Furthermore, the interpretation extends easily to the multivariate case.

The prior sample interpretation of rapid is useful for three reasons. First, standard interpretation tools from regression may be applied to rapid learning parameter and measurement estimates, since any rapid learning results are equivalent to sample-based weighted estimation results. This is important because all such tools are based on sound theory and many of them are powerful. Second, in applications where pertinent prior samples are available, initial parameter values may be computed from available prior records. Resulting rapid learning estimates at any time point will combine information from prior sample records with record information from time point 1 up to the current time. Rapid learning predictions that use prior estimates based on actual data in this way may be more precise, especially during early concurrent operation.

Third, a variety of important applications arise where subjective expertise must be included in decision making, but quantifying expertise in such settings is often difficult [9]. The prior sample interpretation of rapid learning indicates the following approach: the expert may be asked to first construct a prior sample in keeping with his or her expectations. The prior sample may then be used to compute initial parameter values for rapid learning operation. Rapid learning operation may then commence, combining empirical information with subjective prior information from "mind experiments" to produce posterior predictions [4]. (This interpretation is also equivalent to certain subjective interpretations associated with Bayesian statistical inference [3].)

5.2.2 Paired Associate Learning and Prediction

This sub-section extends the above univariate rapid learning results to the bivariate case. Prediction in this case requires associations among pairs of measurements to be learned, hence the term paired associate learning (see section 4.3). The sub-section begins with an overview of paired associate learning concepts, followed by an example, followed by a formulation. Since the same rapid learning concepts apply for the bivariate case as for the univariate case, they are not repeated here (see section 5.2.1). To keep the description manageable, the following results are based on measurements all having plausibility values of 1 (see section 4.2 — the more general case is covered in section 5.3).

As in the bivariate sample-based case (see section 5.1.2), bivariate rapid learning prediction produces estimates of dependent variable values at each time point, based on independent variables at the same time point along with regression estimates from a prior sample. Rapid learning results differ from sample-based results in the bivariate case, along the same lines as the univariate case (see section 5.2.2). In addition to these differences, rapid learning methods routinely estimate each paired associate as a function of the other paired associate, while separate regression functions must be estimated using standard statistical methods. Rapid learning results at each time point thus correspond to two sample-based multiple regression results: one for the first variable and one for the second variable.

As with the univariate case, *Rapid Learner*™ software either supplies initial parameter values by default or utilizes user-supplied parameter values. Default mean values are measurement values from the first record, default variance values are 1, and the default covariance value is 0. A 0 initial covariance value produces for each variable initial prediction functions that depend only on the prior mean for that variable.

If users are supplying their own variance and covariance values, care should be taken to produce values that are achievable from real samples. Otherwise, numerical problems may result. Also, user-supplied initial variance values should be greater than zero, to avoid dividing by zero during regression weight computations.

Figure 5.2.2.1 shows data and tolerance band estimates from a paired associate rapid learning example. Each row in the figure corresponds to one time point from 0 to 14. From time point 1 onward, a measurement record is received containing two measurement values (measured value from points 16 onward are not shown). At each time point, a weight is assigned to each value as well. In addition, at each time point from 1 onward, two predicted measurement values are computed, which are not shown. However, the two tolerance bands that are centered around these predicted values are shown.

Trial	Impact	Learning	Observed Values		Tolerance Bands	
Number	Weight	Weight	1	2	1	2
0	1.0	—	9.7	3.2	—	—
1	1.0	1	7.3	0.1	(7.70, 11.7)	(1.20, 5.20)
2	1.0	1/2	11.0	3.8	(6.20, 13.5)	(-0.56, 8.65)
3	1.0	1/3	8.7	2.2	(5.93, 12.5)	(-1.98, 5.56)
4	1.0	1/4	8.3	0.7	(4.95, 10.7)	(-1.74, 4.84)
5	2.0	2/5	6.8	1.7	(6.16, 11.4)	(-3.13, 2.94)
6	2.0	2/7	10.8	4.1	(7.52, 13.3)	(0.32, 6.60)
7	2.0	2/9	13.3	5.3	(9.24, 14.2)	(3.11, 3.58)
8	2.0	2/11	9.2	2.8	(7.08, 12.4)	(0.15, 5.09)
9	2.0	2/13	14.9	6.1	(11.1, 15.9)	(4.41, 8.89)
10	4.0	2/15	12.6	5.4	(10.7, 15.7)	(2.69, 6.90)
11	4.0	4/19	7.4	2.0	(6.34, 11.0)	(-0.70, 3.31)
12	4.0	4/23	11.5	3.7	(8.23, 13.0)	(2.33, 6.11)
13	4.0	4/27	14.6	5.8	(11.5, 16.1)	(4.25, 7.80)
14	4.0	4/31	8.9	3.1	(7.78, 12.2)	(0.86, 4.16)

FIGURE 5.2.2.1 A Paired Associate Rapid Learning Example (Courtesy of Rapid Clip Neural Systems, Inc.)

The impact weights assigned to trials 0-14 in Figure 5.2.2.1 are the same as the weights assigned to trials 1 through 15 for the previous weighted estimation example (see section 5.1.3). The measurement values for trials 1-14 are the same as measurement values 2-15 for the same previous example.
The results in Figure 5.2.2.1 are based on recursively updating parameter values from time point 1 onward, along with initializing parameter values at time point 0. The initial values in the mean parameter vector $\mu^{\{0\}}$ are set to the first observation values from the previous example (see section 5.1.3). The initial values of the variance and error variance parameters — $v_{11}^{\{0\}}$, $v_{22}^{\{0\}}$, $v_1^{[E]\{0\}}$, and $v_2^{[E]\{0\}}$ are set to 1, and the initial value of the covariance parameter $v_{21}^{\{0\}}$ is set to 0. The tolerance band values at time point 0 reflect those initial mean values along with a tolerance band factor of 2, corresponding to a 95% confidence interval (see section 5.1.1).

When compared with corresponding sample-based results (see section 5.1.3), the rapid learning results of Figure 5.2.2.1 show the same contrasts as in the univariate case, including the following:

- Both predicted values are available from point 1 onward.
- Both predicted values shift toward recently observed values from point 1 onward.
- Both tolerance band values are available from point 1 onward.
- Both tolerance bands shift as the process changes.
- Both tolerance bands become narrower during the changing process.

Figure 5.2.2.2 graphically shows the Figure 5.2.2.1 measurement records, along with three prediction function plots. The vertical scale contains $j_2^{\{t'\}}$ values, the horizontal scale contains $j_1^{\{t'\}}$ values, and each ($j_1^{\{t'\}}$, $j_2^{\{t'\}}$) pair is represented accordingly by a bullet, labeled by its corresponding time point. The three linear plots are functions for predicting $j_2^{\{t'\}}$ values from $j_1^{\{t'\}}$ values at time points 5, 10, and 15. Corresponding functions for predicting $j_1^{\{t'\}}$ from $j_2^{\{t'\}}$ are not shown.

The plots in Figure 5.2.2.1 show the recursive nature of rapid learning prediction. Although only three prediction functions are shown, one such function is available for *each* dependent variable at *each* point from time 1 onward. Each such plot is based on the current value of the other dependent variable, along with learning from a sample that has occurred up to, but not including, the current moment. Although corresponding plots cannot be created for the multivariate case, the bivariate plot should be kept in mind for the multivariate case as well. As in the bivariate case, prediction functions for each dependent variable

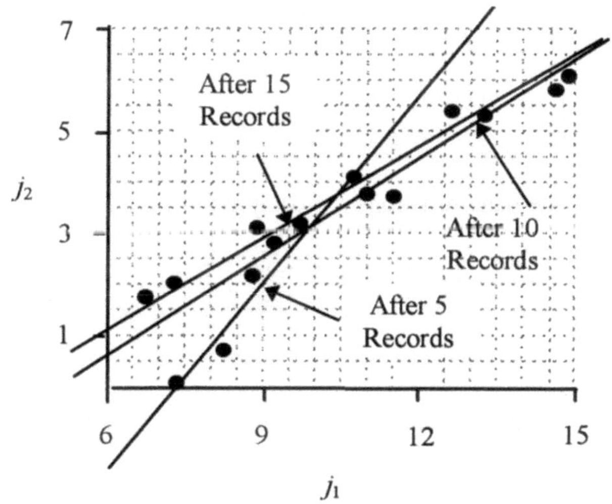

FIGURE 5.2.2.2 A Paired Associate Prediction Example
(Courtesy of Rapid Clip Neural Systems, Inc.)

based on all other variables are updated in this way and available at each time point.

Rapid learning formulas for the bivariate case are given next. At each time point t' ($t' = 1, 2, \ldots$), paired associate learning involves updating the following estimates:

1. The current mean vector, $\mu^{\{t'\}} = (\mu_1^{\{t'\}}, \mu_2^{\{t'\}})$
2. The current variance vector, $(v_{11}^{\{t'\}}, v_{22}^{\{t'\}})$
3. The sample covariance value, $v_{21}^{\{t'\}}$

Before these statistics are updated, two prediction functions are created of the form,

$$\hat{j}_1^{\{t'\}} = \mu_1^{\{t'-1\}} + (j_2^{\{t'\}} - \mu_2^{\{t'-1\}})\rho_{21}^{\{t'-1\}}$$

and

$$\hat{j}_2^{\{t'\}} = \mu_2^{\{t'-1\}} + (j_1^{\{t'\}} - \mu_1^{\{t'-1\}})\rho_{12}^{\{t'-1\}}$$

where the regression weights are computed as follows:

$$\rho_{21}^{\{t'-1\}} = v_{21}^{\{t'-1\}} / v_{11}^{\{t'-1\}}$$

and

$$\rho_{12}^{\{t'-1\}} = v_{21}^{\{t'-1\}} / v_{22}^{\{t'-1\}}.$$

In addition, tolerance band functions are computed of the form,

$$(\underline{j}_1^{\{t'\}}, \bar{j}_1^{\{t'\}}) = (\hat{j}_1^{\{t'\}} - c v_1^{[E]\{t'-1\}1/2}, \hat{j}_1^{\{t'\}} + c v_1^{[E]\{t'-1\}1/2})$$

and

$$(\underline{j}_2^{\{t'\}}, \bar{j}_2^{\{t'\}}) = (\hat{j}_2^{\{t'\}} - c v_2^{[E]\{t'-1\}1/2}, \hat{j}_2^{\{t'\}} + c v_2^{[E]\{t'-1\}1/2}).$$

Once the two prediction functions and the two tolerance bands have been computed, parameter estimate updating begins. First, mean estimates are updated as follows:

$$\mu_1^{\{t'\}} = (l^{\{t'\}} j_1^{\{t'\}} + \mu_1^{\{t'-1\}})/(1 + l^{\{t'\}})$$

and

$$\mu_2^{\{t'\}} = (l^{\{t'\}} j_2^{\{t'\}} + \mu_2^{\{t'-1\}})/(1 + l^{\{t'\}}),$$

where $l^{\{t'\}}$ is computed just as in the univariate case (see section 5.2.1). Next, variance and covariance estimates are updated as follows:

$$v_{11}^{\{t'\}} = [l^{\{t'\}}(j_1^{\{t'\}} - \mu_1^{\{t'\}})^2 + c v_{11}^{\{t'-1\}}]/(1 + l^{\{t'\}}),$$

$$v_{22}^{\{t'\}} = (l^{\{t'\}}(j_2^{\{t'\}} - \mu_2^{\{t'\}})^2 + c v_{22}^{\{t'-1\}})/(1 + l^{\{t'\}}),$$

and

$$v_{22}^{\{t'\}} = [l^{\{t'\}}(j_2^{\{t'\}} - \mu_2^{\{t'\}})^2 + c v_{22}^{\{t'-1\}}]/(1 + l^{\{t'\}}),$$

where c is a positive constant that adjusts for mean updating (see section 5.3.3). Finally, the error variance values are updated as follows:

$$v_1^{[E]\{t'\}} = [l^{\{t'\}}(j_1^{\{t'\}} - \hat{j}_1^{\{t'\}})^2 + v_1^{[E]\{t'-1\}}]/(1 + l^{\{t'\}})$$

and

$$v_2^{[E]\{t'\}} = [l^{\{t'\}}(j_2^{\{t'\}} - \hat{j}_2^{\{t'\}})^2 + v_2^{[E]\{t'-1\}}]/(1+l^{\{t'\}}).$$

5.2.3 Differential Reinforcement Learning

The relationship between rapid learning weights and sample-based impact weights offers a clear interpretation of recursive learning operation. If all measurements have the same impact weight, if the initial sample has the same impact weight, and if the standard rapid learning weight formula is used (see section 5.2.1), straightforward algebra yields the following learning weight schedule:

$$l^{\{t'\}} = 1/t', \; t' = 1, 2, \ldots .$$

The consequences and interpretation of using this learning weight schedule are simple: at any point t' ($t' = 1, 2, \ldots$), prediction is determined by the results of a sample-based regression based on all observations up to but not including that time point, with all observations and the initial sample receiving equal impact.

For cases where all observations have equal impact values but the initial sample has a different impact value, the learning weight schedule has the following form:

$$l^{\{t'\}} = \frac{1}{t' - 1 + w^{\{0\}}/w}, \; t' = 1, 2, \ldots ,$$

where $w^{\{0\}}$ is the initial sample weight and w is the common weight shared by observations at each following time point.

Many other learning schedules are possible as well, including those shown in Figure 5.2.3. The first three schemes shown produce relative impact weights that are *liberal,* in that more recent observations have more impact on learning than less recent observations. Among these, the second schedule from the top is the most liberal, because the last 50 observations have twice the learning impact as the 50 observations before that, and so on. The center schedule is more liberal than the top schedule, because each equal impact block has 4 times the impact as the preceding block in the former case, compared to 2 times the impact in the top case. The shapes of learning impact schedules like these three are governed by two user-supplied parameters (see section 2.1.4): a block width parameter and a block ratio parameter (see section 5.3.3).

Liberal learning may produce more accurate prediction than equal impact learning in applications where relationships change during information processing. For example, equal impact learning was compared with liberal block learning in one case study where Standard & Poors Index values were forecast every minute during one day of trading [10]. The criterion for comparison was how frequent-

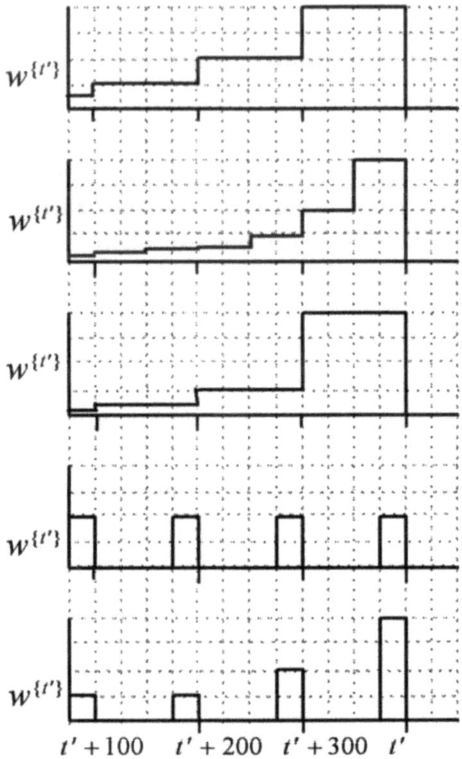

FIGURE 5.2.3 Some Differential Learning Sequences
(Courtesy of Rapid Clip Neural Systems, Inc.)

ly *Rapid Learner*™ software could accurately forecast whether the index would be higher or lower five minutes after each minute during the trading day. When equal impact learning was used, the software successfully forecast increases or decreases five minutes later slightly worse than if such forecasts were made completely randomly. When blocked learning was used, however, the successful forecast rate increased to 68%. The conclusion drawn was that relationships were changing so quickly that liberal learning was needed to keep up with them. (In this study, the block length parameter was set to 30 minutes, and measurements in each block were given twice the impact values as measurements in the preceding block [10].)

Rapid Learner™ software allows the user to choose from a variety of learning schedules, including equal impact learning as well as blocked learning. Alternatively, users may supply learning weight schedules to meet individual specification needs. The two learning weight plots at the bottom of Figure 5.2.3 are examples of user-supplied impact schedules that might be useful. In some applications, prediction functions for monitoring, forecasting, or control may be required only during certain time intervals, for example while machinery to be monitored is operating. In that case, the second plot from the bottom may be appropriate, with positive learning weights assigned only in disjoint blocks. The learning weights that must be provided to produce such a plot are simple to deduce, based on their straightforward impact interpretation (see section 5.2.1). Hybrid impact schedules — such as the bottom plot in the figure that combines interrupted learning with a liberal learning schedule — may easily be imposed as well.

Learning weight schedules may also be set to perform conventional sample-based analysis, in applications where rapid learning methods complement alternative methods (see sections 2.2.3 and 2.3.3). In such applications, sample-based regression results may be obtained using *Rapid Learner*™ software, by imposing equal impact learning on the estimation sample, followed by learning weights that are all 0. Resulting rapid learning prediction functions will be nearly the same as statistical regression functions, insofar as equal impact weights are high, relative to initial impact weights.

5.3 MULTIVARIATE CONCURRENT LEARNING AND PREDICTION

This section introduces the rapid learning counterpart to sample multivariate regression (see section 5.1.3). Only three new issues are described in detail here, because all other concepts and concerns are straightforward extensions to the univariate and bivariate case (see section 5.1).

5.3.1 Solving a Certain Numerical Problem

Beginning with the trivariate case, at every time point a record is received containing 3 measurements, which may be represented by a vector $\boldsymbol{j}^{\{t'\}}$ containing 3 elements. At each time point, each measurement value is predicted as a function of the other 2 measurement values at that time point, along with parameters that have been estimated from a prior sample. In the trivariate case, these parameters include the three means, the three variances, and three covariances — one for each pair of measurements. These parameters are updated in the same way as their bivariate counterparts (see section 5.2.3).

A basic requirement associated with sample-based multivariate regression is that estimation samples must be at least as large as the number of variables per record. Otherwise, numerical problems will occur because parameter estimates require inverting the sample covariance matrix (see section 5.1.5), which will be of deficient rank [7]. Since rapid learning prediction is equivalent to sample-based estimation from prior samples, and since rapid learning begins prediction

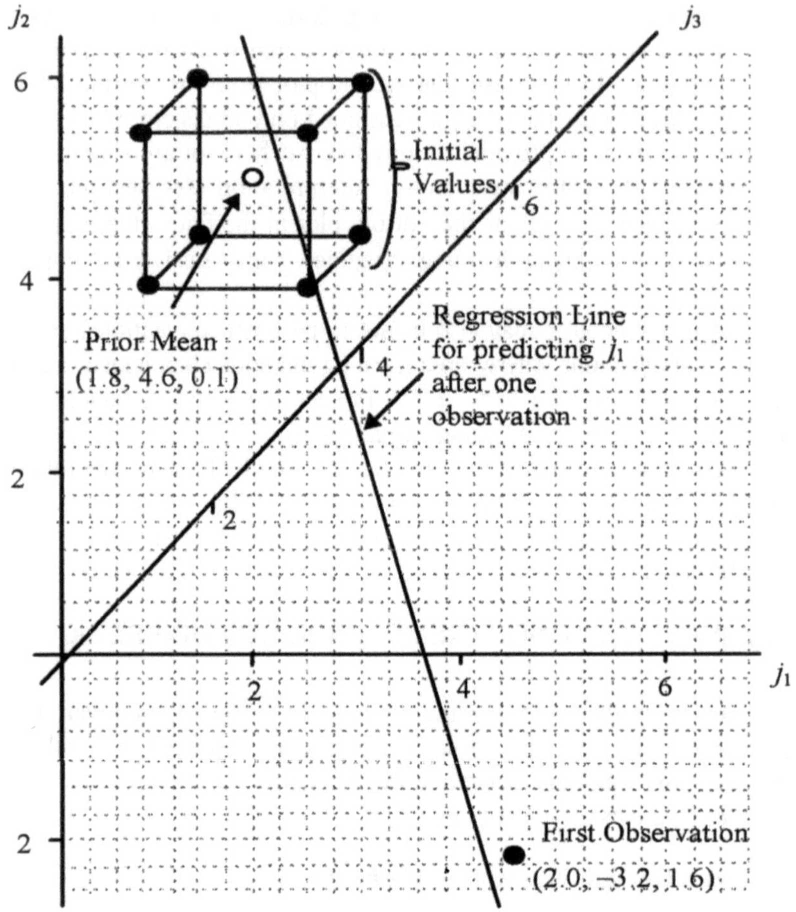

FIGURE 5.3.1 A Trivariate Initialization Example
(Courtesy of Rapid Clip Neural Systems, Inc.)

from trial 1 onward, prior rapid learning samples must satisfy this basic requirement as well.

Rapid learner™ software satisfies the full-rank prior sampling requirement by creating initial parameter values as if they were from satisfactory prior samples. These initial parameter estimates are satisfactory in terms of 3 basic requirements: they solve the numerical problem, they produce an initial mean vector that is the same as the first record measurement vector, and they produce initial prediction functions as if all measurements were initially statistically independent [11]. In the trivariate case, initial parameter estimates are set as if they came from 8 observations having variance values of 1 and covariance values of 0.

(In the more general case involving j^+ measurements, initial parameter estimates are set as if they came from 2^{j^+} measurements, producing an identity covariance matrix.)

Figure 5.3.1 shows prior records graphically for the trivariate case. The first observation is shown by a bullet at the bottom of the figure, the prior mean is shown by a circle, and the 8 prior observations are shown by bullets at the top of the figure. The equation for predicting $\hat{j}_1^{\{2\}}$ from $j_2^{\{2\}}$ and $j_3^{\{2\}}$ after the first observation is shown by the regression line close to the prior mean and the first observation. The other two prediction equations for predicting $\hat{j}_2^{\{2\}}$ and $\hat{j}_3^{\{2\}}$ after one observation are not shown.

The slopes of regression functions, such as the Figure 5.3.1 regression line, depend on the learning weights assigned to records, especially the first few records. If early learning weights are low, then several trials will be required before strong prediction functions begin to emerge. However, low learning weight values will ensure that numerical problems will arise during early learning.

A second numerical problem associated with multivariate regression arises when two or more variables are used that have high linear correlations. *Rapid learner*™ software resolves these problems in much the same way as it satisfies the prior sample requirement: by effectively adding artificial observations that produce variance and covariance values of 1 and 0, respectively. During routine refinement operations, the software checks for indications of such multi-colinearity and adjusts covariances accordingly (see section 8.2).

5.3.2 Learning and Prediction With Missing Measurements

This sub-section deals with settings where some measurements may be missing at a given time point. Two kinds of problems along with rapid learning solutions to them will be introduced: potential estimation as well as prediction distortion due to records having some but not all measurements missing. To keep the description manageable, only the trivariate case is discussed in this sub-section. The more general case is formulated at the end of this chapter (see section 5.3.3).

Suppose that 3 measurements per record are available at each time point, along with a corresponding plausibility vector of length 3 (see section 4.1). In particular, suppose that each element of the plausibility vector is either 0, indicating that its corresponding measurement is missing at that time point, or 1, indicating that its corresponding measurement is not missing at that time point. Suppose further at any time point t' ($t' = 1, 2, \ldots$), mean and regression weights are available from a prior sample up to but not including time t'. In that case, prediction must proceed if 0, 1, 2, or all 3 measurements are missing at time t', based on prior learning up to but not including time t'.

If no measurements are missing as indicated by a plausibility vector of the form $\boldsymbol{p}^{\{t'\}} = (0, 0, 0)$, *Rapid learner*™ software produces estimated measurement values $\widehat{j}_1^{\{t'\}}$ through $\widehat{j}_3^{\{t'\}}$, each of which is a weighted sum of the other two measurement values at that time point (see section 5.3.1). If one or more measurements are missing, the software provides two options: fast prediction and precise prediction. Fast prediction produces estimated measurement values by (a) using prediction functions that have been learned up to time t', and (b) replacing missing independent variables by their means that have been learned up to time t'. The overall effect is to exclude terms from the prediction equation that involve missing variables.

For example, straightforward algebra can show that in the trivariate case, the following prediction equations will result (see section 5.3.3): if all 3 measurement values are missing, each will be predicted by its mean; if measurement 2 is missing as indicated by a plausibility vector of the form $\boldsymbol{p}^{\{t'\}} = (1, 0, 1)$, the predicted measurement value $\widehat{j}_2^{\{t'\}}$ will depend on $j_1^{\{t'\}}$ and $j_3^{\{t'\}}$; the predicted measurement value $\widehat{j}_1^{\{t'\}}$ will depend only on $j_3^{\{t'\}}$; the predicted measurement value $\widehat{j}_3^{\{t'\}}$ will depend only on $j_1^{\{t'\}}$. Similar results hold for other plausibility patterns.

Fast prediction may be highly inaccurate, especially if missing measurements are highly correlated with dependent variable measurements. In missile tracking prediction, for example (see section 1.2), suppose that *Rapid learner*™ software has learned to predict each missile coordinate at each time point t' as a function of a recent history of coordinate values, along with current values of the other 2 coordinates. In that case, regression weights linking current coordinate values to each other may be much higher than regression weights linking current to recent coordinate values. If the same regression weights are used to forecast next missile position, the resulting prediction functions will be inaccurate because *Rapid learner*™ software will have learned to rely on other coordinate values available at the same time. Because of this inaccuracy problem, the software offers alternative *precise prediction* options that are more accurate.

The precise prediction adjustment procedure used by *Rapid learner*™ software replaces regression weights that have been learned from non-missing measurements with regression weights that would have been learned if missing measurements had not been available during learning. (In statistical terms, the software recomputes regression weights based on conditional covariances instead of unconditional covariances — see section 1.3.) The overall effect is a prediction procedure that is precise, but somewhat slower than fast prediction. The loss in speed ranges from a few milliseconds per missing measurement when records each contain 100 measurements or less to about a second per missing measurement when records each contain a few thousand measurements.

A fast and accurate alternative to the above two options for prediction with missing values may be employed in applications where missing values have a known prior structure. In missile tracking control, for example, missing values have a known prior structure in that all three of the next coordinates to be predicted at each time point will always be missing. In this case, a separate prediction may be computed for each next time-point coordinate, each of which is based on current and recent measurements, but not on the other next time-point coordinates. *Rapid learner*™ software routinely computes prediction functions for all forecasting and control operations in this way. This approach may also be advisable in monitoring situations where missing current values will often fit an expected pattern.

Similar concerns and solutions to predicting with missing values hold for learning with missing values. *Rapid learner*™ software deals with missing value learning differently for mean estimation than for covariance estimation. Mean estimation at every time point involves the use of *component learning weights*, with one such weight being computed for each measurement. As in the case for learning with non-missing values (see section 5.2.1), mean estimate updating involves learning weights, prior mean values, and current measurement values. The mean updating formulas are similar, having the following form for updating mean j' ($j' = 1, 2, 3$ in the trivariate case, $j' = 1, 2, \ldots, j^+$ in the general case — see section 5.3.3):

$$\mu_{j'}^{\{t'\}} = (l_{j'}^{[\mathrm{C}]\{t'\}} j_{j'}^{\{t'\}} + \mu_{j'}^{\{t'-1\}})/(1 + l_{j'}^{[\mathrm{C}]\{t'\}}) \;,$$

where the component learning weight $l_{j'}^{[\mathrm{C}]\{t'\}}$ reflects the current learning weight and plausibility value, relative to previous learning weights and plausibility values. The component learning weight is computed as follows:

$$l_{j'}^{[\mathrm{C}]\{t'\}} = w^{\{t'\}} p_{j'}^{\{t'\}} / (w^{\{0\}} + w^{\{1\}} p_{j'}^{\{1\}} + \cdots + w^{\{t'-1\}} p_{j'}^{\{t'-1\}}) .$$

Consequently, if the current plausibility value for a measurement is 0, its mean does not change during updating; but if its plausibility is 1, its value is updated by an amount that factors in the current impact weight, relative to all prior plausible learning weights.

Variance and covariance updating also involves adjusting learning for measurements that may be differentially plausible from trial to trial. Rapid learning design concerns (along with historical storage minimizing requirements) have resulted in formulas of the following form for updating variance j' ($j' = 1, 2, 3$ in the trivariate case, $j' = 1, 2, \ldots, j^+$ in the general case — see section 5.3.3):

$$v_{j'j'}^{\{t'\}} = (l^{\{t'\}} e_{j'}^{\{t'\}2} + c v_{j'j'}^{\{t'-1\}})/(1 + l^{\{t'\}}),$$

where

$$e_{j'}^{\{t'\}} = (j_{j'}^{\{t'\}} - \mu_{j'}^{\{t'\}})p_{j'}^{\{t'\}}$$

and c is a positive constant that adjusts for mean updating (see section 5.3.3). The overall effect is to update variances in the same way as in the non-missing case (see section 5.3.1), except deviance values are set to 0 if their corresponding plausibility values are 0. Formulas for updating covariances between measurements are similar. For example, the covariance between measurement 2 and measurement 1 is updated as follows:

$$v_{21}^{\{t'\}} = (l^{\{t'\}} e_2^{\{t'\}} e_1^{\{t'\}} + c v_{21}^{\{t'-1\}})/(1 + l^{\{t'\}}).$$

Consequently, as in the variance updating case, either deviance value is set to 0 if its plausibility value is 0, resulting in a reduction of the learned covariance value.

The above approach to updating variances and covariances is fast, but it can distort prediction functions if many missing values exist. For example, if one measurement tends to be missing more than another measurement, the first measurement will tend to have a lower estimated variance than the second measurement. Likewise, if one pair of measurements has more missing values than a second pair, the first pair will tend to have a lower estimated covariance than the second pair. As with prediction in the presence of missing values, distinct prediction functions may be learned and utilized if missing values have a known structure to avoid this kind of distortion. For example, suppose that two sets of measurements are involved, one of which may be missing in some cases, the second of which may be missing in other cases, and both of which are missing in still other cases. Under these conditions, learning and predicting three distinct regression systems to deal with the three different cases may be advisable.

For cases where missing values occur more frequently for some variables than for others, *Rapid learner*™ software offers an alternative option that replaces missing values with predicted values from other variables. In that case, each of the above error values $e_{j'}^{\{t'\}}$ becomes replaced with its regression estimate only if it is missing, as follows ($t' = 1, \ldots;\ j' = 1, 2, 3$ in the trivariate case, $j' = 1, 2, \ldots, j^+$ in the general case — see section 5.3.3):

$$(j_{j'}^{\{t'\}} - \mu_{j'}^{\{t'\}})p_{j'}^{\{t'\}} + \hat{j}_{j'}^{\{t'\}}(1 - p_{j'}^{\{t'\}}).$$

It should be noted that direct sample-based counterparts to these missing value options are routinely offered in standard statistical packages. The default option corresponds to replacing missing values with the mean, and the alternative option corresponds to replacing missing values with their regression estimates [8].

The above rapid learning approach for dealing with binary (0 or 1) plausibility values also produces reasonable results when plausibility values between 0 and 1 are assigned. Using such interim plausibility values makes sense in applications where input measurements are actually averages of other measurement values, a proportion of which may be missing from trial to trial, or the number of which may be different from trial to trial. Using interim plausibility values may also be useful in fuzzy control settings [12-13] where subjective plausibility ratings must be assigned to input measurement values.

5.3.3 Basic Concurrent Learning and Prediction Formulas

(This sub-section covers mathematical details that are not essential for practical use of rapid learning methods.)

(The notation in this section follows conventions that are followed throughout the book — see the Glossary.) Concurrent learning and prediction have been formulated to resemble statistical learning and prediction as closely as possible, while being as efficient as possible in terms of speed and storage. The most notable distinction is that rapid learning operation requires that prediction estimates be updated from trial to trial instead of being computed from an estimation sample. This section clarifies the distinction formally by providing rapid learning counterparts to multivariate regression formulas (see section 5.1.5).

The formulation in this section replaces multivariate parameter estimates, which are performed only once per sample, with rapid learning parameter estimates, which are updated recursively. Rapid learning parameter updating functions depend on parameter estimates from the previous point along with concurrent observed measurements.

Consider the case where only one output measurement, say the last element of $\mathbf{j}$, is being predicted from several input values, say all but the last elements of $\mathbf{j}$. This corresponds to the multivariate regression case with $i^+ > 0$ and $d^+ = 1$ (see section 5.1.5). The rapid learning prediction function has the form,

$$\hat{j}_{j^+}^{\{t'\}} = \mu_{j^+}^{\{t'-1\}} + (\mathbf{j}_{(j^+-1)}^{(j^+)\{t'\}} - \boldsymbol{\mu}_{(j^+-1)}^{(j^+)\{t'-1\}})\boldsymbol{\rho}_{(j^+-1\times 1)}^{[j^+|1,\ldots,j^+-1]\{t'-1\}},$$

where the (j^+) symbols in superscripts indicate vectors containing all but the last element of their original counterparts, and the $\{t'-1\}$ superscripts denote parameter estimates that were updated concurrently during the previous trial.

Rapid learning employs recursive *learning weights* $l^{\{t'\}}$ that are linked to impact weights $w^{\{t'\}}$, which may in turn be interpreted as statistical regression record weights (see section 5.1). Rapid learning also employs parameter initializing prior to empirical learning, so that recursive learning can begin at once. Initial parameter impact weights, along with concurrent updating impact weights, provide a basis for computing concurrent learning weights. If an initial impact weight is labeled by $w^{\{0\}}$, then for $t' = 1, \ldots,$ concurrent learning weights are linked to prior impact weights as follows:

$$l^{\{t'\}} = w^{\{t'\}} / \left(w^{\{0\}} + w^{\{1\}} + \cdots + w^{\{t'-1\}} \right).$$

Concurrent learning weights are then applied recursively to update parameter estimates so that rapid learning parameter estimates at each time point will be equivalent to statistical parameter estimates. In particular, straightforward algebra can be used along with the formulas in this section to show that rapid learning estimates at any time point will be the same as weighted statistical estimates based on the prior sample up to, but not including, that time point.

Rapid learning blocked impact schedules are governed by two parameters (see section 5.2.3), one specifying the proportion of overall learning impact for the most recent block after the first block ($w^{[B]}$), and the other specifying the number of time points for each block ($t^{[B]}$). The learning weights for the first block have the form,

$$l^{\{t'\}} = \frac{1}{(1+t')}, \; t' = 1, \ldots, t^{[B]} - 1.$$

The learning weights for subsequent blocks beginning at time points $bt^{[B]}$, ($b = 1, \ldots$) have the form,

$$l^{\{t'\}} = \frac{1}{(w^{[B]} + t')}, \; t' = bt^{[B]}, bt^{[B]} + 1, \ldots, (b+1)t^{[B]} - 1.$$

The recursive mean learning formula has the following simplified form if all prior and concurrent plausibility values are 1: for $t' = 1, \ldots,$ let

$$\boldsymbol{\mu}_{(j^+)}^{\{t'\}} = \left(l^{\{t'\}} \mathbf{j}_{(j^+)}^{\{t'\}} + \boldsymbol{\mu}_{(j^+)}^{\{t'-1\}} \right) / \left(1 + l^{\{t'\}} \right),$$

with

$$\boldsymbol{\mu}_{(j^+)}^{\{0\}} = \mathbf{0}.$$

In the more general case, prior and concurrent plausibility values may vary among component measurements within time points. In this case, mean updating must provide distinct learning weights for each component measurement. Thus,

$$\boldsymbol{\mu}^{\{t'\}}_{(j^+)} = (\boldsymbol{j}^{\{t'\}}_{(j^+)}\widetilde{\mathbf{D}}(\boldsymbol{l}^{[\mathrm{C}]\{t'\}}) + \boldsymbol{\mu}^{\{t'-1\}}_{(j^+)})[\widetilde{\mathbf{D}}(1+\boldsymbol{l}^{[\mathrm{C}]\{t'\}})]^{-1},$$

where each component learning weight has the form,

$$l^{[\mathrm{C}]\{t'\}}_{j'} = w^{\{t'\}}p^{\{t'\}}_{j'} / (w^{\{0\}} + w^{\{1\}}p^{\{1\}}_{j'} + \ldots + w^{\{t'-1\}}p^{\{t'-1\}}_{j'}),$$

with

$$\boldsymbol{p}^{\{0\}}_{(j^+)} = \mathbf{1}.$$

Concurrent regression estimates may be expressed as functions of concurrent covariance matrix elements, just as statistical regression estimates are functions of statistical covariance matrix estimates as in Section 5.1.8. For the case where the last element of $\boldsymbol{j}$ must be predicted from all other elements of $\boldsymbol{j}$, concurrent regression weights satisfy,

$$\boldsymbol{\rho}^{[j^+|1,\ldots,j^+-1]\{t'\}}_{(j^+-1\times 1)} = \boldsymbol{v}^{[11]\{t'\}-1}_{(j^+-1\times j^+-1)}\boldsymbol{v}^{[12]\{t'\}}_{(j^+-1\times 1)},$$

where

$$\boldsymbol{v}^{\{t'\}}_{(j^+\times j^+)} = \begin{bmatrix} \boldsymbol{v}^{[11]\{t'\}}_{(j^+-1\times j^+-1)} & \boldsymbol{v}^{[12]\{t'\}}_{(j^+-1\times 1)} \\ \boldsymbol{v}^{[21]\{t'\}}_{(j^+-1)} & v^{\{t'\}}_{j^+j^+} \end{bmatrix}.$$

Regression weights for predicting any other given element of $\boldsymbol{j}$ from all remaining elements of $\boldsymbol{j}$ may be formulated in a similar way.

Concurrent regression weight learning resembles concurrent mean updating, in that corresponding covariance matrix updating satisfies the same learning weight scheme and is based on the same rationale. The covariance matrix updating formula has the form,

$$\boldsymbol{v}^{\{t'\}}_{(j^+\times j^+)} = (l^{\{t'\}}\boldsymbol{e}^{\{t'\}\mathrm{T}}_{(j^+)}\boldsymbol{e}^{\{t'\}}_{(j^+)} + c\,\boldsymbol{v}^{\{t'-1\}}_{(j^+\times j^+)}) / (1 + l^{\{t'\}}),$$

where $\boldsymbol{e}^{\{t'\}}$ is a current error vector and c is a constant that adjusts for the fact that the previous covariance matrix was not computed using the current mean. For the case where all current plausibility values are 1,

$$e_{(j^+)}^{\{t'\}} = j_{(j^+)}^{\{t'\}} - \mu_{(j^+)}^{\{t'\}}.$$

For the more general case the default *Rapid Learner*™ missing value option is based on

$$e_{j'}^{\{t'\}} = (j_{j'}^{\{t'\}} - \mu_{j'}^{\{t'\}})p_{j'}^{\{t'\}}, \; j' = 1, \ldots, j^+,$$

while the alternative option is based on

$$e_{j'}^{\{t'\}} = (j_{j'}^{\{t'\}} - \mu_{j'}^{\{t'\}})p_{j'}^{\{t'\}} + \hat{j}_{j'}^{\{t'\}}(1 - p_{j'}^{\{t'\}}).$$

Rapid Learner™ software offers one fast and simple option satisfying the above formulation, along with others that are more precise when input plausibility values are zero. The above formulation may not be accurate in cases where several concurrent plausibility values are zero, because it is based on multiple rather than multivariate regression estimates (see section 5.1.3). The more accurate alternative *Rapid Learner*™ options correspond to the above formulation, except input variables with plausibility values of zero are treated as missing dependent variables, in keeping with the proper multivariate formulation. As with other formulations, the formulas used by *Rapid Learner*™ software for precise multivariate prediction are far more efficient than the mathematically equivalent formulas presented here (see section 5.1.3).

Empirical error variance updating has a similar recursive form to mean and covariance updating. The equation for updating error variance element j' has the following form ($j' = 1, \ldots, j^+$):

$$v_{j'}^{[E]\{t'\}} = [l_{j'}^{[C]\{t'\}}(j_{j'}^{\{t'\}} - \hat{j}_{j'}^{\{t'\}})^2 + v_{j'}^{[E]\{t'-1\}}]/(1 + l_{j'}^{[C]\{t'\}}).$$

CONCLUSION

Future Directions

The linear rapid learning operations described in this chapter have been developed to satisfy regression interpretability constraints as well as high speed computing requirements. Since these constraints leave little room for modeling flexibility and future development, major future changes to this basic formulation are not anticipated. On the other hand, extensions to this basic linear model are potentially limitless, pointing in many interesting directions for future study (see chapter 6). In addition, extended interpretations of the linear model may suggest interesting future inference and application directions. Most notably, the prior sample interpretation that is introduced in this chapter, along with the possibility of creating prior samples subjectively from "mind experiments," is potentially valuable in applications where expertise must be converted into prediction equations.

Several simple variants of the linear learning operations described in this chapter could broaden the rapid learning application realm considerably. Most notably, many application areas such as medical informatics are plagued by patient records and questionnaire responses having missing data. New missing data options, which reduce prediction distortion while maintaining rapid learning capability, are well worth future study.

Summary

This chapter introduces basic concurrent learning and prediction in statistical estimation and prediction terms. First, sample-based linear regression is reviewed conceptually, including detailed descriptions of key concepts such as weighted estimation that are employed by rapid learning methods. Second, concurrent learning and prediction are introduced by explaining and interpreting simple rapid learning operations in regression terms. Third, more detailed issues and concepts associated with certain numerical and missing value problems are introduced and explained in the same terms. Finally, areas for future research are suggested, most notably extended interpretations into the subjective inference realm and improved methods for rapidly dealing with missing data.

REFERENCES

1. N.R. Draper & H. Smith, *Applied Regression Analysis*, New York, Wiley, 1966.
2. F. Mosteller & J.W. Tukey, *Data Analysis and Regression*, Addison-Wesley, Reading, MA, 1977.
3. R.J. Jannarone, K.F. Yu, & J.E. Laughlin, "Easy Bayes Estimation for Rasch Type Models," *Psychometrika*, Vol. 55, pp. 449-460, 1990.
4. R.J. Jannarone,"The ABC of Measurement," Unpublished Technical Report, Machine Cognition Laboratory, University of South Carolina, 1994.
5. E.L. Lehmann, *Theory of Point Estimation*, Wiley, New York, 1983.
6. P.J. Bickel & K.A. Doksum, *Mathematical Statistics: Basic Ideas and Selected Topics*, Holden-Day, San Francisco, 1977.
7. F.A. Graybill, *Matrices With Applications in Statistics*, 2nd Edn., Wadsworth, International, Belmont, CA, 1983.
8. SAS Institute, Inc., *SAS Procedures Guide, Version 6.03*, 4th Edn., SAS Institute, Cary, NC, 1988.

9. J. Pearl, *Probabilistic Reasoning in Intelligent Systems: Networks of Plausible Inference*, Morgan Kaufman, San Mateo, CA, 1988.
10. "Commodity Price Forecasting," *RCNS Technical Report Series*, No. APP96-02, Rapid Clip Neural Systems, Inc., Atlanta, GA, 1996.
11. T.W. Anderson, *An Introduction to Multivariate Statistical Analysis*, 2nd Edn., Wiley, New York, 1984.
12. K.J. Aström & T.J. McAvoy, "Intelligent Control: an Overview and Evaluation," in D.A. White & D.A. Sofge (Eds.), *Handbook of Intelligent Control*, Van Nostrand Reinhold, New York, 1992.
13. B. Kosko, *Neural Networks and Fuzzy Systems*, Prentice-Hall, Englewood Cliffs, NJ, 1992.

6

Extended Concurrent Learning and Prediction

	INTRODUCTION	174
6.1.	ADDITIVE MEASUREMENT FUNCTIONS	175
	6.1.1. Mean Feature Functions	176
	6.1.2. Converting Plausibility Values to Viability Values	177
	6.1.3. Using Mean Feature Functions	179
	6.1.4. Mean Feature Function Formulas	182
6.2.	POWERS AND PRODUCTS OF MEASUREMENTS	183
	6.2.1. Computing and Using Product Feature Functions	183
	6.2.2. Product Feature Function Formulas	185
6.3.	HISTORICAL MEASUREMENT FUNCTIONS	187
	6.3.1. Computing and Using Recent History Features	187
	6.3.2. Historical Feature Function Formulas	189
6.4.	SPATIAL PREDICTION	190
	6.4.1. Computing and Using Spatial Features	190
	6.4.2. Spatial Feature Separability	191
6.5.	BINARY AND CATEGORICAL MEASUREMENT FUNCTIONS	193
	6.5.1. Computing and Using Binary and Categorical Features	193
	6.5.2. Binary and Categorical Feature Function Formulas	199
6.6.	COMPOUND FEATURE FUNCTIONS	202
	6.6.1. Powers of Features	202
	6.6.2. Historical Compound Features	207
	CONCLUSION	207
	REFERENCES	208

INTRODUCTION

This chapter extends the linear rapid learning foundations from chapter 5 along several directions, all of which involve converting measurements to measurement features. Measurement features are linear or non-linear measurement functions that are linearly correlated during rapid learning operation. The result of computing and utilizing such features is a rapid learning system that deals with many linear and non-linear applications. In addition, the system is able to operate very quickly because it is able to update linear relationships among such features in real time (see chapter 5).

The chapter begins with simple functions of measurements alone called mean features (section 6.1). The chapter next introduces powers and products of measurements (section 6.2), historical trend functions (section 6.3), and functions of measurements that are organized in space (section 6.4). Historical features capture trends in time, while spatial features capture local trends in space instead of time. The rapid learning system uses a simple and very fast method to represent nonlinear trends in time or space as linear functions of historical measurements and nearest-neighbor measurements, respectively. These measurement feature computing methods allow the rapid learning system to achieve its design goal: very fast learning and prediction over a very general domain of applications. In spatial applications, the rapid learning system can operate especially quickly because spatial learning and prediction operations may be separated and performed in parallel (see section 6.4.2).

Measurement features introduced in this chapter are separated into functions of arithmetic, binary, and categorical input measurements. An arithmetic measurement may have any numerical value, and either the measurement itself or the measurement raised to a power may be linearly correlated with other measurements or measurement features. Arithmetic features introduced in this chapter include means and powers of arithmetic measurements (see sections 6.1 and 6.2). Arithmetic features are also used to describe how the rapid learning system deals with features computed when some dependent measurements are missing (see section 6.1.2).

A binary measurement may only have a value of 0 or 1, and it may be linearly correlated with other measurements or measurement features. Powers of binary measurements are not useful, however, because any power of a binary value is the same binary value (0 raised to any power is 0; 1 raised to any power is 1). A categorical measurement may only have an integer value of 1 through its highest categorical value, which must be 3 or higher (two-category variables are treated as equivalent binary variables by *Rapid Learner*™ software). Categorical measurements are not linearly correlated with other measurements and measurement features directly by *Rapid Learner*™ software. Instead, each of them is routinely converted to a set of equivalent binary features (see section 6.5), which in turn are linearly correlated with other measurements and measurement features.

Compound measurement features described at the end of this chapter combine the individual features introduced earlier in the chapter (see section 6.6). These include powers of features that can solve a variety of practical problems (see section 6.6.1), along with historical features that capture trends among spatial features (see section 6.6.2). Some interesting special cases are introduced including rapid learning counterparts to general linear models and solutions to the parity problem (see section 6.6.1), along with rapid learning versions of shape and motion detection devices (see section 6.6.2).

The results in this chapter combine others that are described earlier in this book. The rapid learning system separates feature computations and other feature operations into modular components (see section 4.1). At each time point, input measurement values are first converted to input feature values by the Input Transducer (see section 4.1.2). Concurrent prediction learning and operations are then performed by the Kernel (see section 4.1.3), producing updated learning along with predicted feature values. Predicted feature values are then converted to predicted measurement values by the Output Transducer (see section 4.1.1). These Kernel operations on feature values are precisely the same basic operations for linear prediction and forecasting as those that have been introduced earlier (see chapter 5).

The feature functions introduced in this chapter, then, are computed and applied in a way that inserts extended Transducer functionality between input measurements and Kernel operations. As a result, all Kernel operations presented in chapter 5 — as if they were used with *measurement* and *plausibility* values— are now presented as they are actually used — with *feature* and *viability* values.

For readers without statistical training, some of the material in this chapter may be difficult to follow — even some material that is not marked as optional. These readers may be forced to either scan the material quickly or cover it slowly.

6.1 ADDITIVE MEASUREMENT FUNCTIONS

This section introduces the creation and use of measurement features that are additive functions of arithmetic measurements. The section describes how *Rapid Learner*™ software deals with the most familiar case: replacing several measurements with their mean. The section also describes issues associated with prediction when some of the measurements being averaged are missing, and it explains how *Rapid Learner*™ software resolves these issues by using mean feature viability functions.

6.1.1 Mean Feature Functions

In structural aircraft testing applications, a test article such as an aircraft wing is equipped with strain gauges to measure load distributions during testing (see section 1.1). These strain gauges may fit into natural clusters, within which correlations among gauges are similar. For example, several strain gauge types are used during testing, some being used on wing beams, others being used on the skin of the wing, and so on. Each type may naturally fit into a cluster, with the property that correlations among two gauges within the same cluster are more similar than correlations among any two gauges from two different clusters. In a similar way, gauges that are about the same distance from each other tend to have similar correlations with each other. Therefore, they may belong in the same cluster.

Often, when several similar measurements belong in a cluster, they may be replaced by a single mean (i.e., average) feature value, for purposes of predicting other measurements. For example, if an entire aircraft is being tested, average strain gauge readings from one wing may be used as a global statistic for predicting strain readings on other parts of the aircraft. To give a more familiar example, items in academic and other mental tests are routinely added to produce global test scores, which are then used to produce test grades and academic success predictions [1-2].

For prediction purposes, mean features are equivalent to sums because they both produce the same correlations for predicting other measurements [1]. Unweighted means or sums are also good approximations to weighted means and sums in many prediction applications [3]. Many powerful methods for dealing with weighted sums and means have been developed within the field of psychometrics [4]. However, since most of these methods require iterative estimation, they are not suitable for rapid learning applications.

Replacing clusters of measurements by their means produces a few rapid learning advantages. First, means tend to predict other measurements or measurement features more precisely than individual measurements in isolation [1-2]. Second, replacing many measurements with a single mean reduces storage requirements considerably by reducing the number of features to be inter-correlated (see section 2.4.2). Using means instead of measurements allows rapid learning problems involving many measurements to be solved on conventional computers, instead of requiring specialized computers with massive memory capacity. Third and most important, replacing many measurements with a single mean allows rapid learning operations to proceed more quickly than they would if the measurements themselves were used (see section 2.4.2). For example, if two measurement clusters each containing 100 measurements were replaced by two averages, rapid learning operations could proceed 10 times more quickly because 10 times fewer computations would be required. Even greater speed improvements are possible if operations can be performed on special-purpose hardware after clustering instead of on conventional computers without clustering (see sections 2.4.1 and 2.4.2).

Rapid Learner™ software identifies measurement clusters for averaging in one of two ways: through user-supplied specifications or through automatic model refinement specifications. *Rapid Learner*™ feature function specifications allow arithmetic and binary measurement clusters to be defined by users (see section 3.6.2). Model refinement operations identify measurement clusters by evaluating the similarity of regression weight profiles for each pair of measurements (see section 8.1). The clustering process results in each arithmetic and binary measurement being placed in one and only one cluster for averaging.

Once each such cluster has been identified, *Rapid Learner*™ software replaces measurement values within each cluster at each time point by their mean. Only means instead of individual measurement values are then utilized for rapid learning operations (see section 6.1.3). Each mean is the proportion of measurements within each cluster that are plausible (see section 5.2). For example, suppose that 50 measurements $j_1^{\{t'\}}$ through $j_{50}^{\{t'\}}$, along with their 50 corresponding plausibility values $p_1^{\{t'\}}$ through $p_{50}^{\{t'\}}$, are to be averaged at time t' ($t' = 1, 2, \ldots$). If the first 30 have plausibility values of 1 and the last 20 have plausibility values of 0, then the resulting mean value will be the average of the first 30. If the first 25 have plausibility values of 0 and the last 25 have plausibility values of 1, the resulting mean value will be the average of the last 25. The more general case where some plausibility values are between 0 and 1 is treated similarly (see section 6.1.3).

6.1.2 Converting Plausibility Values to Viability Values

When measurement values are converted to measurement feature values at each time point, measurement plausibility values must be converted to feature *viability* values at each time point as well (see section 4.1). This sub-section describes such viability functions in general and viability functions for mean features in particular.

Just as every measurement has one plausibility value, which may be any number from 0 to 1 (section 5.3.2), every feature function value has one viability value, which may be any number from 0 to 1. For example, the mean feature function formula presented at the end of the last section produces one such feature value at every point as a function of its 50 independent measurement variable values and its 50 independent plausibility variable values. Likewise, a value between 0 and 1 is obtained at each time point for each mean feature's viability. The viability value is obtained by evaluating an appropriate mean viability function, with its independent values set to corresponding plausibility values at this time point.

Viability value assignment is governed by several concerns, ranging from straightforward to subtle. When independent variable plausibility values are all 0, then the only resulting viability value that makes sense is 0. Likewise, when all component plausibility values are 1, then the only resulting viability value that makes sense is 1. However, when independent variable plausibility values are mixed, or some fall between 0 and 1, the proper approach to computing viability values is less clear. One approach, setting a feature's viability value to the lowest of its component plausibility values, is straightforward computationally. It also results in conservative use of feature functions from a regression viewpoint (see section 6.1.3). However, the approach results in substantial loss of information in some applications. For example, suppose that a mean feature is computed as the average of 250 strain gauge measurements, only one of which is not available at a certain time point. In this case, proceeding as if all other 249 measurements are missing at that time point makes little sense. Computing and utilizing a reduced domain average made up of the other 249 values, along with a less conservative viability estimate, makes more sense.

Mean viability value assignment in the reduced domain case is governed by how the rapid learning system uses viability values during prediction. Once feature and corresponding feature viability values are obtained, they are used for prediction by the rapid learning system in essentially the same way as measurement and viability values are used when features are not involved (see section 5.3.2, chapter 7). If the default *Rapid Learner*™ prediction option is selected, feature values weighted by their corresponding viability values are used to predict other feature and/or measurement values. The overall effect of a viability value less than one, then, is the same as if the feature has a lower regression weight than the value that has been learned up to that time point. This use of viability values makes sense, provided that they are assigned accordingly. In particular, if reducing the domain of a mean produces a reduction in the regression weight between the mean and other measurements by a fixed proportion, then its viability should be the same proportion.

Some guidance for evaluating reduced domain mean viability comes from mental test theory [1], which considers test predictability as a function of test item characteristics. Standard test theory results that give test predictability as a function of reduced test lengths also apply to the viability based on reduced mean domains. Although, even relatively simply cases do not produce simple regression weight reduction formulas (see section 6.1.4), they do show that as the number of items in a test decreases the regression weight between that test score and other measurements decreases as well. Since appropriate regression weight formulas are difficult to implement quickly, *Rapid Learner*™ software computes simplified viability values by default. The default viability value at any time point is simply the average of plausibility values for measurements making up a mean feature (see section 6.1.4).

It should be noted that users may resolve feature viability issues in their own way if they choose to do so, by separating feature function calculations from the rapid learning system. In cases involving a mean feature that is to be linked with other measurements, users may calculate a mean feature and its viability value at each time point, prior to supplying measurements and plausibility values to the rapid learning system. The mean feature and its corresponding viability value may then be supplied to the rapid learning system at that time point as if it were a measurement value and its corresponding plausibility value. At the same time point, other measurement values and their plausibility values may be supplied. The overall effect will be the same as computing the mean feature and its viability internally by the rapid learning system, but users will have full control over viability calculations instead of relying on pre-programmed system procedures.

6.1.3 Using Mean Feature Functions

The rapid learning system has two concurrent operating modes (see section 2.1): real-time prediction and occasional feature refinement. Real-time prediction involves the following steps:

- Converting measurement values to feature values based on feature functions, which may or may not be linear.
- Linearly predicting missing feature values from each other if necessary.
- Using predicted feature values to obtain predicted measurement values if necessary.

The feature functions used by the rapid learning system are governed by user feature specifications (see section 3.6) and automatic refinement operations (see section 8.2). In modular terms (see section 4.1), features are computed by the Input Transducer then their interrelations are learned and utilized by the Kernel. Subsequently, predicted feature values are converted to predicted measurement values by the Output Transducer. Once feature functions have been determined, all rapid learning prediction functions have been determined as well. The following example illustrates a case in which features are used in this way.

Suppose that a rapid learning system is being used with $j^+ = 152$ measurements per record, $j_1^{\{t'\}}$ through $j_{152}^{\{t'\}}$ ($t' = 1, 2, \ldots$). Suppose further that the system produces 4 features per record: one mean feature $m_1^{\{t'\}}$ that is the average of the first 100 measurements, a second mean feature $m_2^{\{t'\}}$ that is the average of the next 50 measurements, and two other features that are simply $m_3^{\{t'\}} = j_{151}^{\{t'\}}$ and $m_4^{\{t'\}} = j_{152}^{\{t'\}}$. In this case the system produces 4 viability values per record as well: two mean viability values $v_1^{\{t'\}}$ and $v_2^{\{t'\}}$ associated with $m_1^{\{t'\}}$ and $m_2^{\{t'\}}$ (see section 6.1.2), and two other viability values that are simply $v_3^{\{t'\}} = p_{151}^{\{t'}$ and $v_4^{\{t'\}} = p_{152}^{\{t'\}}$ (see section 5.3.2). At the beginning of each time point t', then, the Input Transducer converts each input measurement vector $\boldsymbol{j}^{\{t'\}}$ of length $j^+ = 152$ and its corresponding plausibility vector $\boldsymbol{p}^{\{t'\}}$ of the same length to an input feature vector $\boldsymbol{m}^{\{t'\}}$ of length $m^+ = 4$ and its corresponding viability vector $\boldsymbol{v}^{\{t'\}}$ of the same length.

Once the input feature vector $\boldsymbol{m}^{\{t'\}}$ is computed, it is supplied to the Kernel module. The Kernel module then performs linear prediction and learning operations based on $\boldsymbol{m}^{\{t'\}}$, just as if it were performing linear prediction and learning operations with arithmetic measurements (see chapter 5). Once these operations are performed, then, predicted features values $\hat{\boldsymbol{m}}^{\{t'\}}$ are available for a variety of monitoring, forecasting, and control operations. These predicted features are also used as predicted output measurement values in place of missing input measurement values, if necessary. In cases where feature values are identical to measurement values, such as with $m_3^{\{t'\}} = j_{151}^{\{t'\}}$ and $m_4^{\{t'\}} = j_{152}^{\{t'\}}$ in above example, output measurement values are produced just as when no features are involved (see section 5.2.3). That is,

$$\hat{j}_{151}^{\{t'\}} = m_3^{\{t'\}} p_{151}^{\{t'\}} + \hat{m}_3^{\{t'\}} (1 - p_{151}^{\{t'\}})$$

and

$$\hat{j}_{152}^{\{t'\}} = m_4^{\{t'\}} p_{152}^{\{t'\}} + \hat{m}_4^{\{t'\}} (1 - p_{152}^{\{t'\}}).$$

In cases involving mean features, such as $m_1^{\{t'\}}$ and $m_2^{\{t'\}}$ in the previous example, output measurement values are produced similarly, namely,

$$\hat{j}_{j'}^{\{t'\}} = j_{j'}^{\{t'\}} p_{j'}^{\{t'\}} + \hat{m}_1^{\{t'\}} (1 - p_{j'}^{\{t'\}}), \ j' = 1, \ldots, 100$$

and

$$\hat{j}_{j'}^{\{t'\}} = j_{j'}^{\{t'\}} p_{j'}^{\{t'\}} + \hat{m}_2^{\{t'\}} (1 - p_{j'}^{\{t'\}}), \ j' = 101, \ldots, 150.$$

In this way, measurements that are missing and associated with a mean feature may be replaced with the mean feature's estimated value. This estimated value may depend on measurements associated with the same mean feature that are not missing, or on predicted values, or on both.

6.1.4 Mean Feature Function Formulas

(This section covers mathematical details that are not essential for practical use of rapid learning methods.)

(The notation in this section follows conventions that are followed throughout the book — see the Glossary.) Given a vector of arithmetic or binary measurements,

$$\boldsymbol{j}_{(j^+)}^{\{t'\}},$$

and their corresponding plausibility vector,

$$\boldsymbol{p}_{(j^+)}^{\{t'\}},$$

their arithmetic mean feature is computed as follows:

$$m^{[\mathrm{FM}]\{t'\}} = (p_1^{\{t'\}} j_1^{\{t'\}} + \cdots + p_{j^+}^{\{t'\}} j_{j^+}^{\{t'\}}) / c^{\{t'\}},\ c^{\{t'\}} > 0,$$

$$= 0,\ c^{\{t'\}} = 0,$$

where

$$c^{\{t'\}} = p_1^{\{t'\}} + \cdots + p_{j^+}^{\{t'\}}.$$

Since the default *Rapid Learner*™ prediction option effectively reduces regression weights by factoring in viability values (see section 5.3.3), mean feature viability values would ideally be set to proportionate regression weight reductions due to missing component measurements. To give one special case from mental test theory [1], consider the following situation:

- A test is made up of j^+ items and the mean test score x is being used to predict another variable y.
- The variable y and all items have variance values of 1.
- Every item used to compute x has the same correlation coefficient χ (see chapter 7) with all other items.
- Each item used to compute x has the same correlation coefficient χ with y.

In this case, the regression weight for predicting the dependent variable y from the mean (or total) test score x is

$$\rho^{[Y|X]} = \frac{j^+\chi}{[j^+ + j^+(j^+ - 1)\chi]^{1/2}}.$$

Now suppose that a mean feature $x^{\{t''\}}$ is computed by the rapid learning system at every time point t'' in the same way ($t'' = 1, \ldots, t' - 1$), and that $\rho^{[Y|X]\{t'-1\}}$ is available for predicting $y^{\{t'\}}$ ($t' = 2, \ldots$). Suppose further that $\rho^{[Y|X]\{t'-1\}}$ has been learned from means without missing values. Finally, suppose that $c^{\{t'\}}$ items making up $x^{\{t'\}}$ are missing at the current time point ($c^{\{t'\}} = 1, \ldots, j^+$). Then the appropriate regression coefficient for predicting $y^{\{t'\}}$ from $x^{\{t'\}}$ is

$$\nu^{\{t'\}}\rho^{[Y|X]\{t'-1\}},$$

where

$$\nu^{\{t'\}} = \frac{c^{\{t'\}}\chi}{[c^{\{t'\}} + c^{\{t'\}}(c^{\{t'\}} - 1)\chi]^{1/2}} \bigg/ \frac{j^+\chi}{[j^+ + j^+(j^+ - 1)\chi]^{1/2}}.$$

This precise value for a viability coefficient is difficult to implement even in the above idealized case, because computing it requires maintaining an accurate estimate of the correlation coefficient χ between component measurements. In the more general case, computing precise viability coefficients is far more difficult, because all inter-measurement correlations must be maintained. For this reason, *Rapid Learner*™ software computes the following simplified viability coefficient by default:

$$\nu^{\{t'\}} = c^{\{t'\}}/j^+.$$

Once a mean feature value $m^{[FM]\{t'\}}$ is computed in this way, it is supplied to the rapid learning Kernel function just as if it were any other arithmetic measurement (see chapter 5), resulting in an output estimated feature value $\widehat{m}^{[FM]\{t'\}}$. This value is then used to estimate measurement values as follows:

$$\widehat{j}_{j'}^{\{t'\}} = j_{j'}^{\{t'\}}p_{j'}^{\{t'\}} + \widehat{m}^{[FM]\{t'\}}(1 - p_{j'}^{\{t'\}}),\ j' = 1, \ldots, j^+.$$

6.2 PRODUCTS AND POWERS OF MEASUREMENTS

This section introduces product features of arithmetic variables. Product features based on binary variables, categorical features and other features are described later in this chapter (see section 6.5).

6.2.1 Computing and Using Product Feature Functions

Product features for an arithmetic variable are computed simply by raising that variable to an integer-valued power. Users specify power features by providing a power degree $\bar{p}$ ($\bar{p}$ = 1, 2, . . .). Once the power degree is specified, $\bar{p}$ power features for an arithmetic variable $j^{\{t'\}}$ are created. For example, if $\bar{p}$ = 5 is specified to establish fifth-order features, then feature values of the form,

$$\boldsymbol{m}^{[\mathrm{FP}]\{t'\}} = (j^{\{t'\}2}, j^{\{t'\}3}, j^{\{t'\}4}, j^{\{t'\}5}),$$

are computed at every time point t' (t' = 1, 2, . . .).

When a rapid learning model involves several arithmetic measurements and when power features are specified at a given power level $\bar{p}$, power features up to $\bar{p}$ are computed for each of the measurements in the model. In addition, cross-products among the measurements are computed from degree 2 up to $\bar{p}$. For example, if three measurements are involved and a third-order model is specified, then power features have the form,

$$\boldsymbol{m}^{[\mathrm{FP}]\{t'\}} = (j_1^{\{t'\}2}, j_1^{\{t'\}}j_2^{\{t'\}}, j_1^{\{t'\}}j_3^{\{t'\}}, j_2^{\{t'\}2}, j_2^{\{t'\}}j_3^{\{t'\}}, j_3^{\{t'\}2}, j_1^{\{t'\}3}, j_1^{\{t'\}2}j_2^{\{t'\}},$$

$$j_1^{\{t'\}2}j_3^{\{t'\}}, j_1^{\{t'\}}j_2^{\{t'\}2}, j_1^{\{t'\}}j_2^{\{t'\}}j_3^{\{t'\}}, j_1^{\{t'\}}j_3^{\{t'\}2}, j_2^{\{t'\}3}, j_2^{\{t'\}2}j_3^{\{t'\}}, j_2^{\{t'\}}j_3^{\{t'\}2}, j_3^{\{t'\}3}).$$

Each power feature is used to predict other features and/or measurements within a rapid learning system, but not the measurement that was used to create it. Using a power of a measurement $j^{\{t'\}}$ to predict the measurement statistically makes no sense, because $j^{\{t'\}}$ can always be computed precisely by evaluating an appropriate function. For example, $j^{\{t'\}}$ can be computed as the square root of $j^{\{t'\}2}$ or as the cube root of $j^{\{t'\}3}$, without the need for prediction from other variables. Using powers of some measurements to predict others does make sense, however. For example, strain values may be related to load values squared, textile colors may be periodic functions of their values at other positions on a two-dimensional grid, and so on.

The rapid learning system uses a measurement's powers to predict other measurements but not to predict the measurement itself. The system restricts the use of powers in this way by creating a dedicated model for each measurement variable and including features accordingly. In the above case, three models are required, one for predicting each of the three measurements from each of the other measurements and their powers. The features associated with the three models are as follows:

$$\boldsymbol{m}_1^{[\mathrm{FP}]\{t'\}} = (j_2^{\{t'\}},\ j_3^{\{t'\}},\ j_2^{\{t'\}2},\ j_2^{\{t'\}} j_3^{\{t'\}},\ j_3^{\{t'\}2},$$

$$j_2^{\{t'\}3}, j_2^{\{t'\}2} j_3^{\{t'\}},\ j_2^{\{t'\}} j_3^{\{t'\}2},\ j_3^{\{t'\}3},\ j_1^{\{t'\}}),$$

$$\boldsymbol{m}_2^{[\mathrm{FP}]\{t'\}} = (j_1^{\{t'\}},\ j_3^{\{t'\}},\ j_1^{\{t'\}2},\ j_1^{\{t'\}} j_3^{\{t'\}},\ j_3^{\{t'\}2},$$

$$j_1^{\{t'\}3}, j_1^{\{t'\}2} j_3^{\{t'\}},\ j_1^{\{t'\}} j_3^{\{t'\}2},\ j_3^{\{t'\}3},\ j_2^{\{t'\}}),$$

and

$$\boldsymbol{m}_3^{[\mathrm{FP}]\{t'\}} = (j_1^{\{t'\}},\ j_2^{\{t'\}},\ j_1^{\{t'\}2},\ j_1^{\{t'\}} j_2^{\{t'\}},\ j_2^{\{t'\}2},$$

$$j_1^{\{t'\}3}, j_1^{\{t'\}2} j_2^{\{t'\}},\ j_1^{\{t'\}} j_2^{\{t'\}2},\ j_2^{\{t'\}3},\ j_3^{\{t'\}}).$$

In this way, no power of a measurement appears in the same prediction model for that measurement.

As with mean features and other features, each power feature value at each time point has a corresponding viability value for prediction at that time point. The viability value for a measurement raised to a power is simply the same as the viability value for the measurement. The viability value involving any cross-product among measurements is the product among the plausibility values for the measurements involved in the cross-product. In cases where all plausibility values are either 0 or 1, then, the cross-product viability will be 1 unless any of the component plausibility values are 0. In other cases when plausibility values are 0, the cross-product viability will be 0. In still other cases where plausibility values are between 0 and 1, the product method for computing cross-product viability values will also be between 0 and 1, in keeping with a certain independence assumption (see section 6.2.2).

Once arithmetic measurements are predicted from other measurement values and their power feature values, their predicted values are used for monitoring, forecasting, control, and replacing their input missing values. Uses of predicted measurements based on powers are essentially the same as uses of the arithmetic measurements alone (see sections 5.2 and 5.3).

Initializing covariance values among power features is more complicated than initializing covariance values among measurements because power features tend to have high covariances with each other. For example, suppose that 21 measurements have the values (−10, −9, . . ., −1, 0, 1, 2, . . .,10). Suppose further that each of these measurements is placed in a record, along with their features of order 2 through 6. When their covariances are computed using standard formulas (see chapter 5), these measurements and their three odd-order powers have positive covariance values, while their three even-order powers also have positive covariance values. Initializing to reflect these positive covariances is important since early rapid learning performance will otherwise be distorted if initial values reflect unrealistic zero-valued covariances among powers.

Rapid Learner™ software computes initial covariance values among power values based on actual samples on a grid along the lines of the above example, along with their power values. The records in these samples are spaced on the grid and weighted so that the measurement values themselves in any initialization sample will have a mean of 0 and a variance of 1. This produces initial power values on the same scale as initial measurement values. Weights that approximate those coming from a normally distributed sample are computed. Instead of normal weights, however, triangular distribution weights are used as approximations (see section 6.2.2). This avoids numerical difficulties when $\bar{p}$ is large.

6.2.2 Product Feature Function Formulas

(This section covers mathematical details that are not essential for practical use of rapid learning methods.)

(The notation in this section is consistent with other book conventions — see the Glossary.) The following trend feature function formulas are mathematically equivalent to those used by *Rapid Learner*™ software. However, the formulas used by the software produce power values much more quickly, and they produce high-order power values much more accurately.

Product feature values for an arithmetic variable are computed by raising values for that variable to an integer-valued power. Users specify power features by providing a power degree $\bar{p}$ ($\bar{p} = 1, \ldots$). Once the power degree is specified, $\bar{p}$ power features for an arithmetic variable $j^{\{t'\}}$ are created of the form,

$$\boldsymbol{m}^{[\mathrm{FP}]\{t'\}} = \left(j^{\{t'\}2}, j^{\{t'\}3}, \ldots, j^{\{t'\}\bar{p}}\right).$$

When several product features are involved, cross-products up to the power $\bar{p}$ are utilized as well. For example, if j^+ feature values are involved, then second-order power features include,

$$\boldsymbol{m}^{[\mathrm{FP}]\{t'\}} = (j_1^{\{t'\}2}, j_1^{\{t'\}} j_2^{\{t'\}}, \ldots, j_1^{\{t'\}} j_{j^+}^{\{t'\}}, j_2^{\{t'\}2}, \ldots, j_{j^+-1}^{\{t'\}} j_{j^+}^{\{t'\}}, j_{j^+}^{\{t'\}2}).$$

Basic higher-order feature functions for arithmetic variables are computed similarly.

Rapid Learner™ power feature initializing utilizes covariances among powers for each measurement from an initializing sample. The initializing sample size is $\bar{p}+1$, sample values are equally spaced about 0, and sample weights are assigned according to a discrete triangular probability distribution [5]. Triangular distributions approximate normal distributions in the continuous case [6], which in turn is approximated in the discrete case when $\bar{p}$ is large. Discrete triangular grid distributions that are used by the rapid learning system have been derived to have mean values of 0 and variance values of 1. When $\bar{p}$ is even, grid values have the form

$$g_{(\bar{p}+1)} = \frac{1}{2}[(\bar{p})(\bar{p}+2)/6]^{1/2}(-\bar{p}/2, -\bar{p}/2+1, \ldots, \bar{p}/2),$$

and for $g' = 1, \ldots, \bar{p}/2$, corresponding weight values have the form

$$w_{g'} = w_{\bar{p}-g'} = c_1 g'$$

while

$$w_{1+\bar{p}/2} = c_1(1+\bar{p}/2),$$

where

$$c_1 = 4/(\bar{p}+2)^2.$$

When $\bar{p}$ is odd, grid values have the form

$$g_{(\bar{p}+1)} = \frac{1}{2}[(\bar{p}+3)(\bar{p}+5)/6]^{1/2}(-(\bar{p}+1)/2, -(\bar{p}+1)/2+1, \ldots, (\bar{p}+1)/2),$$

and for $g' = 1, \ldots, (\bar{p}+1)/2$, corresponding weight values have the form,

$$w_{g'} = w_{\bar{p}-g'+2} = c_0 g',$$

where

$$c_0 = 4/(\bar{p}+1)(\bar{p}+3).$$

Viability values for power features are computed as follows: when a power feature is a measurement raised to a power, the viability for the feature is simply the same as the viability for the measurement. When a power feature involves cross-products among measurement values, its viability value is the product of the viability values for the measurement values involved in the cross-product. The rationale for computing cross-product viability values in this way is based on assuming that each measurement value is an average among several "quantum" measurement values, and each measurement plausibility value is the proportion of such quantum measurements that are missing [7]. If it is further assumed that missing quantum values are distributed independently between measurements involved in a given cross-product, it follows that the expected proportion of quantum cross-products involving non-missing quantum values will be the product of viability values involved. The same rationale holds when powers among mean features are used instead of powers among arithmetic measurements, provided that viability values for mean features are computed according to the *Rapid Learner*™ default (see section 6.1.4).

6.3 HISTORICAL MEASUREMENT FUNCTIONS

This section introduces measurement functions derived from recently observed measurement values. Such values are stored at every time point in a recent feature memory (see section 4.1.2), so that the necessary feature functions may be computed. The feature functions may then be correlated with current as well as future measurement values to increase prediction precision.

6.3.1 Computing and Using Recent History Features

Recent history features are functions of measurements that have been observed in the recent past. Recent history features are used to predict current feature values and to forecast future history values based on recent trends (see chapter 8). The rapid learning system computes recent history feature values at each time point as functions of recent measurement values, according to two time parameters: a time step size parameter $h^{[\mathrm{STEP}]}$ and a number of time steps parameter $h^{[\mathrm{STEPS}]}$. The most recent history measurement used is one time point in the past, the next most recent measurement is $1+h^{[\mathrm{STEP}]}$ time points in the past, and so on.

For example, if an $h^{[\mathrm{STEPS}]}$ value of 5 is specified, then recent history features will be based on 5 past time point measurements, and if an $h^{[\mathrm{STEP}]}$ value of 3 is specified, these will be separated by 3 time points. Measurements going back 13 time points will then be used to compute recent history values as follows: at time point $t' = 100$, the 5 recent values will be

$$\boldsymbol{j}^{[\mathrm{H}]\{100\}} = (j^{\{99\}}, j^{\{96\}}, j^{\{93\}}, j^{\{90\}}, j^{\{87\}}),$$

at time point $t' = 101$ the 5 recent values will be

$$\boldsymbol{j}^{[\mathrm{H}]\{101\}} = (j^{\{100\}}, j^{\{97\}}, j^{\{94\}}, j^{\{91\}}, j^{\{88\}}),$$

and so on. In this way, the rapid learning system uses a "moving window" of recent measurement values to predict current and future measurement values.

The rapid learning system uses a recent history of measurement values for prediction by converting them to recent history *trend* feature values. Trend feature values are computed by fitting polynomials to recent history measurement values. Polynomial degrees are specified by the *Rapid Learner*™ software user. For example, suppose that the user specifies $h^{[\mathrm{STEPS}]} = 6$ recent history time points, an $h^{[\mathrm{STEP}]} = 1$ time point separation between each of them, and a $\overline{p} = 2$ degree (quadratic) polynomial. In this case, at time point $t' = 100$, a second-degree polynomial will be fit to the 6 values,

$$\boldsymbol{j}^{[\mathrm{H}]\{100\}} = (j^{\{99\}}, j^{\{98\}}, j^{\{97\}}, j^{\{96\}}, j^{\{95\}}, j^{\{94\}}),$$

at time point $t' = 101$, a second-degree polynomial will be fit to the 6 values,

$$\boldsymbol{j}^{[\mathrm{H}]\{101\}} = (j^{\{100\}}, j^{\{97\}}, j^{\{94\}}, j^{\{91\}}, j^{\{88\}}),$$

and so on.

Trend feature values at each time point are coefficients of polynomials that fit the recent history measurement values at that time point. For the above example, at time point $t' = 100$, the rapid learning system computes a feature vector of the form,

$$\boldsymbol{m}^{[\mathrm{H}]\{100\}} = (m_1^{[\mathrm{H}]\{100\}}, m_2^{[\mathrm{H}]\{100\}}, m_3^{[\mathrm{H}]\{100\}}).$$

The first element in the trend feature vector is the constant coefficient for a second-degree polynomial fit to the 6 points in $\boldsymbol{j}^{[\mathrm{H}]\{100\}}$, the second element in the vector is the linear coefficient for the second-degree polynomial, and the third element is the quadratic coefficient for the second-degree polynomial. Likewise, at time point $t' = 101$, the rapid learning system computes a feature vector of the form,

$$\boldsymbol{m}^{[H]\{101\}} = (m_1^{[H]\{101\}}, m_2^{[H]\{101\}}, m_3^{[H]\{101\}}).$$

The elements of this feature vector are second-degree polynomial coefficients that fit the recent history measurement vector at time point $t' = 101$. In this way, the rapid learning system computes recent history feature values at every time point t' (once enough time has elapsed to create the necessary initial recent measurement history — $t' = h^{[STEP]}(h^{[STEPS]} - 1) + 2, h^{[STEP]}(h^{[STEPS]} - 1) + 3, \ldots$).

Once recent history trend feature values are computed in this way, they are correlated with current measurements during rapid learning operations. After these correlations have been learned, they are used to predict current and future measurement values. For example, second-order trends were used to predict missile coordinates in a case study (see section 1.3.1). The rapid learning system learned to use these second-order trends quite effectively. In one instance, the system learned that the second-order polynomial coefficient based on a recent history of vertical coordinate values is positively correlated with the next-position vertical coordinate. That is, recent history measurements showing a concave upward trend predict next-position values that are higher than recent history measurements showing a concave downward trend. The rapid learning system correlates other trends with dependent variables and uses such learned correlations in a similar way.

In this way, the rapid learning system correlates trend feature values *linearly*, in order to produce *nonlinear* prediction functions in the time domain. As the missile tracking example graphically shows (see Figure 1.3.1), this method can predict highly nonlinear functions of time. At the same time, the method can keep up with measurements arriving very quickly (see section 2.4.1).

Rapid Learner™ software treats missing values for trend feature computations in a simple way, which allows fast operation when many trend coefficients are computed. The software simply replaces each missing historical $j^{\{t'\}}$ measurement value by its corresponding estimated $\hat{j}^{\{t'\}}$ value, which was obtained at a previous time point when the historical measurement at the current time was a current measurement at that time.

6.3.2 Historical Feature Function Formulas

(This section covers mathematical details that are not essential for practical use of rapid learning methods.)

(The notation in this section uses conventions that are used throughout the book — see the Glossary.) The following trend feature function formula is mathematically equivalent to those used by *Rapid Learner*™ software. However, the formulas used by the software produce trend values much more quickly.

For any pre-specified power value $\overline{p}$ ($\overline{p} = 2, \ldots$), the basic historical trend feature function formula is

$$\boldsymbol{m}_{(\overline{p}+1)}^{[\mathrm{H}]\{t'\}} = \boldsymbol{j}^{[\mathrm{H}]\{t'\}} \boldsymbol{p} (\boldsymbol{p}^{\mathrm{T}} \boldsymbol{p})^{-1},$$

where

$$\boldsymbol{j}_{(h^{[\mathrm{STEPS}]})}^{[\mathrm{H}]\{t'\}} = (j^{\{t' - h^{[\mathrm{STEP}]}(h^{[\mathrm{STEPS}]}-1)-1\}}, \ldots, j^{\{t'-1-h^{[\mathrm{STEP}]}\}}, j^{\{t'-1\}}$$

and

$$\boldsymbol{p}_{(h^{[\mathrm{STEPS}]} \times \overline{p}+1)} = \begin{bmatrix} 1 & 1 & \cdots & 1 \\ 1 & 2 & \cdots & 2^{\overline{p}} \\ \vdots & \vdots & \ddots & \vdots \\ 1 & h^{[\mathrm{STEPS}]} & \cdots & h^{[\mathrm{STEPS}]\overline{p}} \end{bmatrix}.$$

6.4 SPATIAL PREDICTION

This section introduces spatial features, which capture trends over space of visual patterns on a grid. Once such trends are captured by spatial features, they are used to predict nearby grid color and light intensity values. Related applications include textile monitoring and visual pattern recognition. Section 6.4.2 outlines ways that spatial features capture trends in space. Spatial rapid learning features are potentially useful in a variety of image processing applications where compact memory concerns are important [8-9], along with rapid learning concerns [10]. Section 6.4.2 outlines how rapid learning Kernel modules may be configured to satisfy both concerns.

6.4.1 Computing and Using Spatial Features

Spatial features closely resemble historical features. Just as historical features correlate trends over time with other features and/or measurements, spatial features correlate trends over space with other features and/or measurements. Just as historical features compute trends over time by fitting polynomials to recent measurements, spatial features compute trends over space by fitting polynomials to nearest neighbor measurements. Just as historical features are based on a "moving window" in time, spatial features are based on a moving window in space. Finally, just as historical features correlate nonlinear temporal trends with current and future outcome variables, spatial features correlated nonlinear spatial trends with local and nearby outcome variables.

The most natural application for spatial features is visual pattern monitoring [8-10]. For example, a textile pattern may be monitored for irregularities by correlating patch intensity values at a given point with patch intensity trends over space from nearest neighbor patches. Insofar as the rapid learning system identifies such correlations from flawless textile patches during prior learning, the system will identify flawed patches during concurrent monitoring operation (see section 8.1). Many of the standard rapid learning system advantages are especially useful in this application because new textile patterns are often used routinely in textile plants. Since the rapid learning system can identify spatial trend correlations and use them for monitoring quickly, effective monitoring of each new textile batch can begin immediately, with no prior learning. Related image compression and smoothing applications are currently under study as well [10].

6.4.2 Spatial Feature Separability

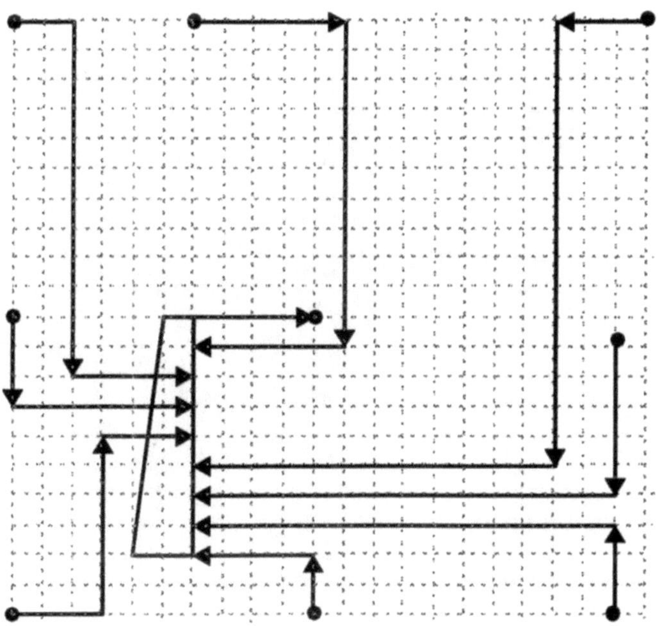

FIGURE 6.4.2.1 A Pattern Completion Kernel
(Courtesy of Rapid Clip Neural Systems, Inc.)

Figure 6.4.2.1 shows how nearest-neighbor pattern completion might be performed by a rapid learning Kernel hardware module (see section 4.1.2). In this case, the Kernel module learns to predict intensity values for the pixel in the center of the figure as a function of intensity values for the eight surrounding pixel values. As with other rapid learning applications, the module updates learning at each time point, and it also predicts the pixel value in the center at each time point if it is missing.

The nearest neighbor configuration in Figure 6.4.2.1 may be easily extended to include several nearest neighbors in space, as well as recent trends in time, along with conjunctive pixel functions. (Historical and conjunctive pixel functions are straightforward extensions of standard compound feature functions computed by *Rapid Learner*™ software — see section 6.5.2.) In this way, a variety of pattern recognition operations may be performed by the Kernel, along with a variety of motion detection operations over time. These configurations and operations resemble others that have received considerable study [8]. However, when used with the Kernel module, they have the added capacity to learn in real time.

Figure 6.4.2.2 shows how an array of Kernel modules can be arranged so that rapid learning and pattern completion operations may proceed for each pixel in a grid, at the same time and in parallel. (To keep the figure simple, connec-

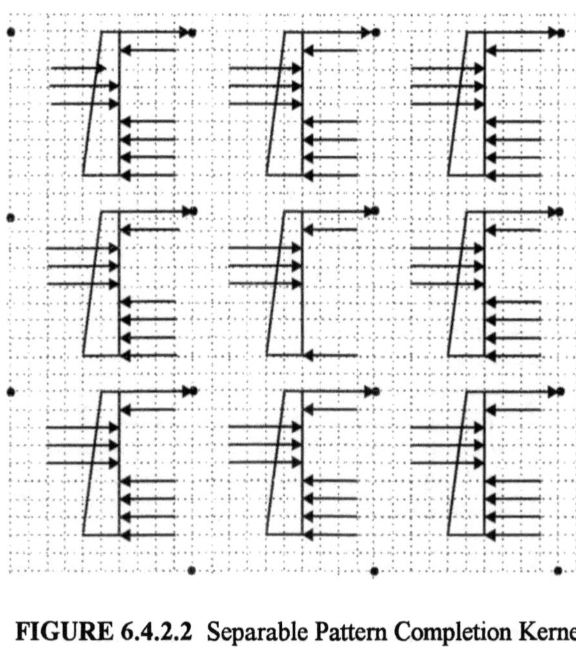

FIGURE 6.4.2.2 Separable Pattern Completion Kernels
(Courtesy of Rapid Clip Neural Systems, Inc.)

tions from each kernel to all nearest neighbor pixels are only partly shown, but such connections are arranged precisely as in Figure 6.4.2.1.) Because they may be arranged in a separable way, these Kernel modules perform pattern completion *and learning* operations very rapidly (see section 2.4.1). In addition, other Kernel modules may be used to predict pixels and pixel features at different points on a grid, at different angles, and for a variety of spatial step sizes. In this way, location-, scale-, and rotation-invariant pattern completion and learning may be achieved in real time.

6.5 BINARY AND CATEGORICAL MEASUREMENT FUNCTIONS

This section introduces binary and categorical feature functions, which allow the rapid learning system to operate quickly and compactly with non-arithmetic measurements. The section begins with detailed examples (see section 6.5.1), and it ends with a general formulation (see section 6.5.2).

6.5.1 Computing and Using Binary and Categorical Features

Binary and categorical feature functions deal with discrete measurements. Binary measurement values may be viewed as labels for one of two logical states or membership in one of two groups (the rapid learning system represents two-category variables as binary measurements). Categorical measurement values may be viewed as membership in one of three or more groups. In rapid learning terms, at every point in time t' ($t' = 1, 2, \ldots$), the rapid learning system may receive a record containing any number of binary measurement values $j^{[B]\{t'\}}$, categorical measurement values $j^{[C]\{t'\}}$, and arithmetic measurement values $j^{[A]\{t'\}}$. Each binary measurement value $j^{[B]\{t'\}}$ may be 0 or 1 and each categorical measurement value $j^{[C]\{t'\}}$ may be an integer from 1 to its maximum number of categories $\bar{c}$ ($\bar{c} = 3, 4, \ldots$). In contrast, each arithmetic measurement $j^{[A]\{t'\}}$ is not discrete, in that it may have any numerical value.

Just as the rapid learning system predicts and learns from arithmetic measurement values at each time point (see chapter 5), it also predicts and learns from binary and categorical measurement values at each time point. The system performs these functions for variables by using binary and categorical features, along the same lines as it uses arithmetic features (see sections 6.1-6.4). The system first converts binary and categorical measurement values to corresponding input feature values. It next uses these input feature values to predict output feature values and update learned inter-feature relations. It then uses the predicted output feature values for monitoring, forecasting, and control, as well as for predicting output binary and categorical measurement values if necessary.

Binary feature functions are relatively simple. Binary input feature values $m^{[B]\{t'\}}$ are exactly the same as their corresponding input measurement values $j^{[B]\{t'\}}$. Once these input feature values are received by the Kernel, they are predicted from other arithmetic, binary, and categorical features to produce output binary feature values $\widehat{m}^{[B]\{t'\}}$. Like all other output feature values, these Kernel output values are arithmetic (i.e., they may take on any numerical value). When an input binary measurement value is missing, the rapid learning system uses its predicted arithmetic feature value to produce its predicted binary feature value according to the following simple formula:

$$\widehat{j}^{[B]\{t'\}} = 1$$

if $\widehat{m}^{[B]\{t'\}} > 0.5$ and

$$\widehat{j}^{[B]\{t'\}} = 0$$

otherwise.

When an input value may or may not be missing as indicated by its corresponding plausibility value (see section 5.2), the rationale for predicting arithmetic measurements and features (see section 6.1) also applies for predicting binary measurements. The function based on that rationale is,

$$\widehat{j}^{[B]\{t'\}} = 1$$

if

$$j^{[B]\{t'\}} p^{\{t'\}} + \widehat{m}^{[B]\{t'\}} (1 - p^{\{t'\}}) > 0.5$$

and

$$\widehat{j}^{[B]\{t'\}} = 0$$

otherwise. The resulting predicted value will be the input value if the plausibility value is 1 and the predicted value if the plausibility value is 0. If the plausibility is between 0 and 1, the resulting predicted value will remain binary, but its criterion will be interpolated between the input value and the predicted value according to its plausibility value.

It should be noted that predicted binary measurement feature values may not necessarily range between 0 and 1. For example, suppose that the status of a machine is represented by a binary variable with a value of 1 if the machine is operating properly and a value of 0 if the machine is not. Suppose further that an arithmetic variable represents ambient temperature around the machine, which normally ranges from near freezing to room temperature. Finally, suppose the rapid learning system has learned that the machine usually operates properly at room temperature, but it operates poorly at the freezing point. In that case, the predicted binary value will be about 1 at room temperature and about 0 at the freezing point.

Continuing the example, suppose temperature suddenly rises to the boiling point. In that case, the predicted binary value will be well above 1. Likewise, if temperature suddenly drops well below freezing, the predicted binary value will be well below 0. In the high temperature case, the rapid learning system will predict that the machine will not break down, in keeping with the above prediction formulas. Likewise, in the low temperature case, it will predict that the machine will break down. The actual binary feature values will be outside the range of 0 and 1, however, because measurements occurred well outside the range of prior learning for the system. *Accordingly, such predicted binary values outside the 0 to 1 range and conclusions based on them should viewed with caution.* This is one of many cases where careful monitoring provided by the rapid learning system may be useful (see section 7.1).

Categorical feature functions are more complicated than binary feature functions. Each categorical variable represents several categories, each pair of which may be contrasted with each other using a binary feature. As a result, several bases may be used for representing categorical variables. The problem of representing several categorical variables in a linear model has a long history of study [11], which has produced a variety of satisfactory solutions for rapid learning applications [12-13]. The particular solution employed by the rapid learning system represents each categorical measurement value $j^{[C]\{t'\}}$ by a corresponding binary feature vector value $\boldsymbol{m}^{[C]\{t'\}}$ at each time point t' ($t' = 1, 2, \ldots$). Each such vector has one less element than the number of categories for its corresponding variable. For example, if a categorical measurement $j^{[C]\{t'\}}$ represents 4 categories, then its corresponding vector $\boldsymbol{m}^{[C]\{t'\}}$ has three elements.

Each possible value for a categorical measurement produces a unique set of binary values in its corresponding feature vector. For example, if $j^{[C]\{t'\}}$ represents 4 categories, then its possible input values from 1 to 4 produce the following corresponding input $\boldsymbol{m}^{[C]\{t'\}}$ values:

$$\boldsymbol{m}^{[C]\{t'\}} = (1, 0, 0)$$

when $j^{[C]\{t'\}} = 1$,

$$\boldsymbol{m}^{[C]\{t'\}} = (0, 1, 0)$$

when $j^{[C]\{t'\}} = 2$,

$$\boldsymbol{m}^{[C]\{t'\}} = (0, 0, 1)$$

when $j^{[C]\{t'\}} = 3$, and

$$\boldsymbol{m}^{[C]\{t'\}} = (0, 0, 0)$$

when $j^{[C]\{t'\}} = 4$.

Once input categorical feature values are computed from input categorical measurement values, they are supplied to the rapid learning Kernel module (see section 4.1.3), in the same way as binary feature values. At each time point t' ($t' = 1, 2, \ldots$), the Kernel updates learning based on these and other feature values it receives, and it supplies predicted categorical feature values $\hat{\boldsymbol{m}}^{[C]\{t'\}}$ in *arithmetic* form. These values are then supplied to the Output Transducer (see section 4.1.1), which converts them to output predicted measurement values $\hat{j}^{[C]\{t'\}}$ in *categorical* form.

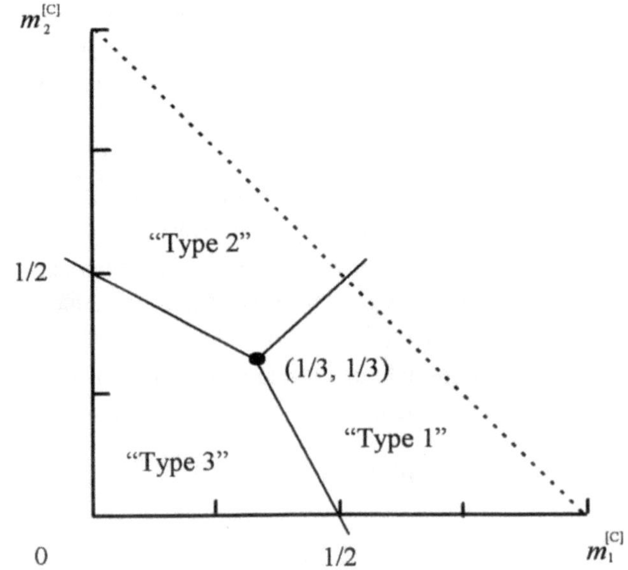

FIGURE 6.5.1.1 Three-Category Decisions
(Courtesy of Rapid Clip Neural Systems, Inc.)

Figure 6.5.1.1 illustrates how the Output Transducer converts output categorical feature values to output category values, in the cases involving three categories. When three categories are involved two features $\hat{m}_1^{[C]\{t'\}}$ and $\hat{m}_2^{[C]\{t'\}}$ are computed, as shown in the figure by $m_1^{[C]}$ and $m_2^{[C]}$, respectively. The figure shows that if $m_1^{[C]}$ is sufficiently large, then "Type 1" is selected, that is $\hat{j}^{[C]\{t'\}}$ is set to 1. Likewise, the figure shows that if $m_2^{[C]}$ is sufficiently large, $\hat{j}^{[C]\{t'\}}$ is set to 2. In addition, the figure shows that if both $m_1^{[C]}$ and $m_2^{[C]}$ are small, $\hat{j}^{[C]\{t'\}}$ is set to 3.

In mathematical terms, the categorical assignment case shown in Figure 6.5.1.1 is expressed as follows:

$$\hat{j}^{[C]\{t'\}} = 2$$

when

$$\hat{m}_2^{[C]\{t'\}} > \hat{m}_3^{[C]\{t'\}}$$

and

$$\hat{m}_2^{[C]\{t'\}} > 1 - (\hat{m}_2^{[C]\{t'\}} + \hat{m}_3^{[C]\{t'\}})\,;$$

$$\hat{j}^{[C]\{t'\}} = 3$$

when

$$\hat{m}_3^{[C]\{t'\}} > \hat{m}_2^{[C]\{t'\}}$$

and

$$\hat{m}_3^{[C]\{t'\}} > 1 - (\hat{m}_2^{[C]\{t'\}} + \hat{m}_3^{[C]\{t'\}})\,;$$

and

$$\hat{j}^{[C]\{t'\}} = 1$$

otherwise.

Figure 6.5.1.2 illustrates how the Output Transducer converts output categorical feature values to output category values, in the cases involving four categories. When four categories are involved three features $\hat{m}_1^{[C]\{t'\}}$, $\hat{m}_2^{[C]\{t'\}}$, and

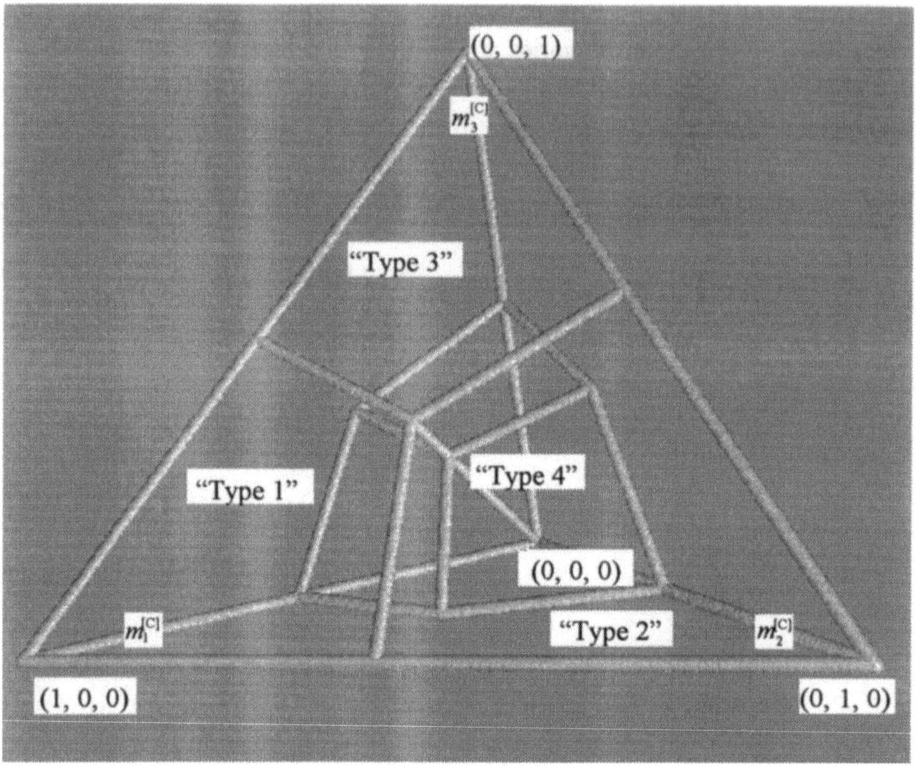

FIGURE 6.5.1.2 Four-Category Decisions
(Courtesy of Rapid Clip Neural Systems, Inc.)

$\hat{m}_3^{[C]\{t'\}}$ are computed, as shown in the figure by $m_1^{[C]}$, $m_2^{[C]}$, and $m_3^{[C]}$, respectively. The figure shows that if $m_1^{[C]}$ is sufficiently large, then "Type 1" is selected, that is $\hat{j}^{[C]\{t'\}}$ is set to 1. Likewise, the figure shows that if $m_2^{[C]}$ or $m_3^{[C]}$ are sufficiently large, $\hat{j}^{[C]\{t'\}}$ is set to 2 or 3, respectively. In addition, the figure shows that if $m_1^{[C]}$, $m_2^{[C]}$ and $m_3^{[C]}$ are small, then $\hat{j}^{[C]\{t'\}}$ is set to 4. The mathematical counterpart to this case is a natural extension to the three-category case (see section 6.5.2).

As in the binary case, it should be noted that predicted binary category variables may fall outside the 0 to 1 range. In terms of Figure 6.5.1.1, two feature coordinates for the three-category case will fall outside the triangle represented by the horizontal axis, the vertical axis, and the dashed line. In terms of Figure 6.5.1.2, the three coordinates for the four-category case will fall outside the pyramid defined by the lines connecting the points (1, 0, 0), (0, 1, 0), (0, 0, 1), and (0, 0, 0). In practical terms, the same conclusions apply for categorical measurements as for binary measurements. Binary category values outside the 0 to 1 range at any given time point indicate that measurement values have occurred at that time point outside the prior learning range. *Accordingly, such values and conclusions based on them should viewed with caution.* As in the binary case, this is another instance when careful monitoring provided by the rapid learning system may be useful (see section 7.1).

In both the binary and the categorical case, the rapid learning system applies viability values for each feature value that are simply the same as the plausibility values for their corresponding measurement value. In this way, each binary and categorical measurement value may be assigned a fuzzy index, indicating strength of membership of the corresponding record in the corresponding category [see section 5.2.2]. In addition, each category associated with a binary or categorical measurement has a more global fuzzy index, namely its category membership probability. The rapid learning system allows users to establish initial category membership probability values, by supplying initial learned parameters to the system (see section 6.5.2). For a binary variable, its initial mean represents its prior probability value. For a categorical variable, the initial mean values of its binary category vector elements represents its prior probability value. For example, if the three binary elements for a four-category variable are assigned initial mean values of (0.2, 0.3. 0.1), then categories 1 through 4 will have prior probability values of 0.2, 0.3, 0.1, and 0.4 (0.4 = 1 – [0.2+0.3+0.1]). By specifying prior probability values in this way, users may establish rapid learning counterparts to fuzzy set membership values.

6.5.2 Binary and Categorical Feature Function Formulas

(This section covers mathematical details that are not essential for practical use of rapid learning methods.)

(The notation in this section uses conventions that are used throughout the book — see the Glossary.) The rapid learning system receives a measurement vector at every time point t' (t' = 1, . . .), of the form,

$$\dot{\boldsymbol{j}}^{\{t'\}}_{(j^+)} = (\dot{\boldsymbol{j}}^{[A]\{t'\}}_{(j^{[A]+})}, \dot{\boldsymbol{j}}^{[B]\{t'\}}_{(j^{[B]+})}, \dot{\boldsymbol{j}}^{[C]\{t'\}}_{(j^{[C]+})}),$$

along with its corresponding plausibility vector,

$$\boldsymbol{p}^{\{t'\}}_{(j^+)} = (\boldsymbol{p}^{[A]\{t'\}}_{(j^{[A]+})}, \boldsymbol{p}^{[B]\{t'\}}_{(j^{[B]+})}, \boldsymbol{p}^{[C]\{t'\}}_{(j^{[C]+})}).$$

The Input Transducer then routinely computes the following Kernel input feature values,

$$\boldsymbol{m}_{(m^{+})}^{\{t'\}} = (\boldsymbol{j}_{(j^{[A]+})}^{[A]\{t'\}}, \boldsymbol{j}_{(j^{[B]+})}^{[B]\{t'\}}, \boldsymbol{m}_{(m^{[C]+})}^{[C]\{t'\}}),$$

along with their corresponding plausibility viability values,

$$\boldsymbol{v}_{(j^{+})}^{\{t'\}} = (\boldsymbol{v}_{(j^{[A]+})}^{[A]\{t'\}}, \boldsymbol{v}_{(j^{[B]+})}^{[B]\{t'\}}, \boldsymbol{v}_{(j^{[B]+})}^{[C]\{t'\}}),$$

such that every arithmetic, binary, and categorical feature viability value is the same as its corresponding measurement plausibility value. (These input feature and corresponding viability values may optionally be augmented with power features, historical features, and so on, along with their own appropriate viability values — see other sections in this chapter.)

The Kernel then computes predicted arithmetic, binary, and categorical feature values of the form,

$$\widehat{\boldsymbol{m}}_{(m^{+})}^{\{t'\}} = (\widehat{\boldsymbol{m}}_{(j^{[A]+})}^{[A]\{t'\}}, \widehat{\boldsymbol{m}}_{(j^{[B]+})}^{[B]\{t'\}}, \widehat{\boldsymbol{m}}_{(m^{[C]+})}^{[C]\{t'\}}).$$

The Output Transducer finally computes predicted arithmetic, binary, and categorical measurement values of the form,

$$\widehat{\boldsymbol{j}}_{(j^{+})}^{\{t'\}} = (\widehat{\boldsymbol{j}}_{(j^{[A]+})}^{[A]\{t'\}}, \widehat{\boldsymbol{j}}_{(j^{[B]+})}^{[B]\{t'\}}, \widehat{\boldsymbol{j}}_{(j^{[B]+})}^{[C]\{t'\}}).$$

Selected details associated with each of these steps are provided next.

Beginning with binary details, *Rapid Learner*™ software sets initial mean values $\mu^{[B]\{0\}}$ for binary features to 0.5 by default, but users may supply alternative $\mu^{[B]\{0\}}$ values to represent real or imagined prior information. In either case, the software sets initial binary feature covariance values to

$$v^{[B]\{0\}} = \mu^{[B]\{0\}}(1 - \mu^{[B]\{0\}}).$$

Binary output measurement values have the form,

$$\widehat{j}^{[B]\{t'\}} = 1, \; j^{[B]\{t'\}}p^{\{t'\}} + \widehat{m}^{[B]\{t'\}}(1 - p^{\{t'\}}) > 0.5,$$

$$= 0 \text{ otherwise.}$$

Turning next to categorical details, if a categorical input variable $j^{[C]\{t'\}}$ at time t' has $\bar{c}$ possible values, its corresponding input feature vector $\boldsymbol{m}^{[C]\{t'\}}$ has the form,

$$\boldsymbol{m}_{(\bar{c}-1)}^{[\mathrm{C}]\{t'\}} = (1, 0, \ldots, 0),\ j^{[\mathrm{C}]\{t'\}} = 1,$$

$$\boldsymbol{m}_{(\bar{c}-1)}^{[\mathrm{C}]\{t'\}} = (0, 1, 0, \ldots, 0),\ j^{[\mathrm{C}]\{t'\}} = 2,$$

$$\vdots$$

$$\boldsymbol{m}_{(\bar{c}-1)}^{[\mathrm{C}]\{t'\}} = (0, \ldots, 0, 1),\ j^{[\mathrm{C}]\{t'\}} = \bar{c}-1,$$

and

$$\boldsymbol{m}_{(\bar{c}-1)}^{[\mathrm{C}]\{t'\}} = (0, \ldots, 0),\ j^{[\mathrm{C}]\{t'\}} = \bar{c}.$$

Predicted categorical measurement values are obtained from predicted categorical feature values by setting

$$\widehat{j}^{[\mathrm{C}]\{t'\}} = 1,\ \max\{\breve{m}_1^{[\mathrm{C}]\{t'\}}, \ldots, \breve{m}_{\bar{c}-1}^{[\mathrm{C}]\{t'\}}, 1-(\breve{m}_1^{[\mathrm{C}]\{t'\}} + \cdots + \breve{m}_{\bar{c}-1}^{[\mathrm{C}]\{t'\}})\} = \breve{m}_1^{[\mathrm{C}]\{t'\}},$$

$$\vdots$$

$$\widehat{j}^{[\mathrm{C}]\{t'\}} = \bar{c}-1,\ \max\{\breve{m}_1^{[\mathrm{C}]\{t'\}}, \ldots, \breve{m}_{\bar{c}-1}^{[\mathrm{C}]\{t'\}}, 1-(\breve{m}_1^{[\mathrm{C}]\{t'\}} + \cdots + \breve{m}_{\bar{c}-1}^{[\mathrm{C}]\{t'\}})\} = \breve{m}_{\bar{c}-1}^{[\mathrm{C}]\{t'\}},$$

and

$$\widehat{j}^{[\mathrm{C}]\{t'\}} = \bar{c} \text{ otherwise,}$$

for each categorical measurement value $j^{[\mathrm{C}]\{t'\}}$, where

$$\breve{m}_{c'}^{[\mathrm{C}]\{t'\}} = m_{c'}^{[\mathrm{C}]\{t'\}} p^{[\mathrm{C}]\{t'\}} + \widehat{m}_{c'}^{[\mathrm{C}]\{t'\}}(1-p^{[\mathrm{C}]\{t'\}}),\ c' = 1, \ldots, \bar{c}-1,$$

and $p^{[\mathrm{C}]\{t'}$ is the plausibility value for $j^{[\mathrm{C}]\{t'\}}$.

Rapid Learner™ software sets initial mean values $\mu^{[\mathrm{C}]\{0\}}$ for binary category features having $\bar{c}$ categories to

$$\boldsymbol{\mu}_{(\bar{c}-1)}^{[\mathrm{C}]\{0\}} = (1/\bar{c}, \ldots, 1/\bar{c})$$

by default, but users may supply alternative $\mu^{[C]\{0\}}$ values to represent actual or imagined prior information. In either case, the software sets initial binary feature covariance values to

$$V^{[C]\{0\}}_{(\bar{c}-1\times\bar{c}-1)} = \widetilde{\mathbf{D}}(\mu^{[C]\{0\}}) - \mu^{[C]\{0\}\mathrm{T}}\mu^{[C]\{0\}}.$$

6.6 COMPOUND FEATURE FUNCTIONS

This section describes feature function combinations, which may combine power and historical features with any variety of other features. Compound power features described in section 6.6.1 extend the rapid learning realm to include a variety of general linear models as special cases, along with straightforward solutions to some difficult problems. Compound historical features described in section 6.6.2 extend the rapid learning realm to include binary and categorical trends over time as well as spatial trends over time. These extensions are only a few among the many possible combinations of rapid learning features, corresponding models, and corresponding applications.

6.6.1 Powers of Features

The rapid learning system computes powers of arithmetic, binary, and categorical measurement in a way that naturally extends power operations with arithmetic measurements only (see section 6.2). Unlike arithmetic measurement powers and cross-products, however, the system restricts powers and cross-products of binary and categorical measurements to avoid redundant power features. Without such restrictions, redundancy problems would arise because binary measurements raised to any positive power are the same as the binary measurements themselves (0 raised to any positive power is 0 and 1 raised to any power is 1). This problem applies to binary as well as categorical features, because input binary as well as categorical features are binary (see section 6.5).

The rapid learning system creates power features involving binary and categorical variables by computing all possible power features up to the user-specified degree $\bar{p}$ ($\bar{p} = 2, 3, \ldots$), subject to the above restrictions. This means that in many cases, powers and cross-products up to degree $\bar{p}$ may be computed for arithmetic variables involved, but not for binary and categorical variables involved. For example, suppose five measurement values arrive at every time point t' ($t' = 1, 2, \ldots$), including two arithmetic values, one binary value, and two categorical values from 1 to 3 and 1 to 4, respectively. This results in eight input feature values (see section 6.5): three corresponding to the three arithmetic and binary measurement values as they stand, two binary category feature values representing values of the first categorical variable from 1 to 3, and three binary category feature values representing values of the second categorical variable from 1 to 4. In rapid learning terms (see section 6.5.2), the feature values have the following form:

$$m^{\{t'\}} = (m_1^{[A]\{t'\}},\ m_2^{[A]\{t'\}},\ m^{[B]\{t'\}},\ m_{11}^{[C]\{t'\}},\ m_{12}^{[C]\{t'\}},\ m_{21}^{[C]\{t'\}},\ m_{22}^{[C]\{t'\}},\ m_{23}^{[C]\{t'\}}).$$

When a second order model is specified in the above case, the rapid learning system uses the above first-order features in addition to the following power features:

$$(m_1^{[A]\{t'\}2},\ m_1^{[A]\{t'\}}m_2^{[A]\{t'\}},\ m_2^{[A]\{t'\}2},$$

$$m_1^{[A]\{t'\}}m^{[B]\{t'\}},\ m_2^{[A]\{t'\}}m^{[B]\{t'\}},$$

$$m_1^{[A]\{t'\}}m_{11}^{[C]\{t'\}},\ m_1^{[A]\{t'\}}m_{12}^{[C]\{t'\}},\ m_1^{[A]\{t'\}}m_{21}^{[C]\{t'\}},\ m_1^{[A]\{t'\}}m_{22}^{[C]\{t'\}},\ m_1^{[A]\{t'\}}m_{23}^{[C]\{t'\}},$$

$$m_2^{[A]\{t'\}}m_{11}^{[C]\{t'\}},\ m_2^{[A]\{t'\}}m_{12}^{[C]\{t'\}},\ m_2^{[A]\{t'\}}m_{21}^{[C]\{t'\}},\ m_2^{[A]\{t'\}}m_{22}^{[C]\{t'\}},\ m_2^{[A]\{t'\}}m_{23}^{[C]\{t'\}},$$

$$m^{[B]\{t'\}}m_{11}^{[C]\{t'\}},\ m^{[B]\{t'\}}m_{12}^{[C]\{t'\}},$$

$$m^{[B]\{t'\}}m_{21}^{[C]\{t'\}},\ m^{[B]\{t'\}}m_{22}^{[C]\{t'\}},\ m^{[B]\{t'\}}m_{23}^{[C]\{t'\}},$$

$$m_{11}^{[C]\{t'\}}m_{21}^{[C]\{t'\}},\ m_{11}^{[C]\{t'\}}m_{22}^{[C]\{t'\}},\ m_{11}^{[C]\{t'\}}m_{23}^{[C]\{t'\}},\ m_{12}^{[C]\{t'\}}m_{21}^{[C]\{t'\}},\ m_{12}^{[C]\{t'\}}m_{22}^{[C]\{t'\}},\ m_{12}^{[C]\{t'\}}m_{23}^{[C]\{t'\}}).$$

Recalling that using a power feature to compute its corresponding measurement makes no sense (see section 6.2.1), the concern applies for using power features to predict arithmetic, binary, and categorical measurement values. When applied to the above example, the rapid learning system uses the following power features (along with other first-order measurement features) to predict first-order measurement features. For predicting the first arithmetic feature,

$$(m_2^{[A]\{t'\}2},\ m_2^{[A]\{t'\}}m^{[B]\{t'\}},\ m_2^{[A]\{t'\}}m_{11}^{[C]\{t'\}},\ m_2^{[A]\{t'\}}m_{12}^{[C]\{t'\}},$$

$$m_2^{[A]\{t'\}}m_{21}^{[C]\{t'\}},\ m_2^{[A]\{t'\}}m_{22}^{[C]\{t'\}},\ m_2^{[A]\{t'\}}m_{23}^{[C]\{t'\}},$$

$$m^{[B]\{t'\}}m_{11}^{[C]\{t'\}},\ m^{[B]\{t'\}}m_{12}^{[C]\{t'\}},\ m^{[B]\{t'\}}m_{21}^{[C]\{t'\}},\ m^{[B]\{t'\}}m_{22}^{[C]\{t'\}},\ m^{[B]\{t'\}}m_{23}^{[C]\{t'\}},$$

$$m_{11}^{[C]\{t'\}}m_{21}^{[C]\{t'\}},\ m_{11}^{[C]\{t'\}}m_{22}^{[C]\{t'\}},\ m_{11}^{[C]\{t'\}}m_{23}^{[C]\{t'\}},\ m_{12}^{[C]\{t'\}}m_{21}^{[C]\{t'\}},\ m_{12}^{[C]\{t'\}}m_{22}^{[C]\{t'\}},\ m_{12}^{[C]\{t'\}}m_{23}^{[C]\{t'\}})$$

are used (similar power features are used for predicting the second arithmetic feature. For predicting the binary feature,

$$m_1^{[A]\{t'\}2},\ m_1^{[A]\{t'\}}m_2^{[A]\{t'\}},\ m_2^{[A]\{t'\}2},$$

$$m_1^{[A]\{t'\}}m_{11}^{[C]\{t'\}},\ m_1^{[A]\{t'\}}m_{12}^{[C]\{t'\}},\ m_1^{[A]\{t'\}}m_{21}^{[C]\{t'\}},\ m_1^{[A]\{t'\}}m_{22}^{[C]\{t'\}},\ m_1^{[A]\{t'\}}m_{23}^{[C]\{t'\}},$$

$$m_2^{[A]\{t'\}}m_{11}^{[C]\{t'\}},\ m_2^{[A]\{t'\}}m_{12}^{[C]\{t'\}},\ m_2^{[A]\{t'\}}m_{21}^{[C]\{t'\}},\ m_2^{[A]\{t'\}}m_{22}^{[C]\{t'\}},\ m_2^{[A]\{t'\}}m_{23}^{[C]\{t'\}},$$

$$m_{11}^{[C]\{t'\}}m_{21}^{[C]\{t'\}},\ m_{11}^{[C]\{t'\}}m_{22}^{[C]\{t'\}},\ m_{11}^{[C]\{t'\}}m_{23}^{[C]\{t'\}},\ m_{12}^{[C]\{t'\}}m_{21}^{[C]\{t'\}},\ m_{12}^{[C]\{t'\}}m_{22}^{[C]\{t'\}},\ m_{12}^{[C]\{t'\}}m_{23}^{[C]\{t'\}})$$

are used. For predicting the first categorical variable features,

$$(m_1^{[A]\{t'\}2},\ m_1^{[A]\{t'\}}m_2^{[A]\{t'\}},\ m_2^{[A]\{t'\}2},$$

$$m_1^{[A]\{t'\}}m^{[B]\{t'\}},\ m_2^{[A]\{t'\}}m^{[B]\{t'\}},$$

$$m_1^{[A]\{t'\}}m_{21}^{[C]\{t'\}},\ m_1^{[A]\{t'\}}m_{22}^{[C]\{t'\}},\ m_1^{[A]\{t'\}}m_{23}^{[C]\{t'\}},\ m_2^{[A]\{t'\}}m_{21}^{[C]\{t'\}},\ m_2^{[A]\{t'\}}m_{22}^{[C]\{t'\}},\ m_2^{[A]\{t'\}}m_{23}^{[C]\{t'\}},$$

$$m^{[B]\{t'\}}m_{21}^{[C]\{t'\}},\ m^{[B]\{t'\}}m_{22}^{[C]\{t'\}},\ m^{[B]\{t'\}}m_{23}^{[C]\{t'\}})$$

are used (similar power features are used for predicting the second category features).

In this way, at each time point several prediction operations are proceeding at once, each of which is based on its own prediction model. Just as linear rapid learning prediction models correspond to standard regression models (see chapter 5), the prediction models in the above example correspond to <u>general linear models</u> [13], including the following:

- Predicting the first arithmetic variable from the second arithmetic variable corresponds to a non-linear <u>regression</u> analysis — this form of prediction occurs at time points when all binary and categorical plausibility values are 0.
- Predicting both arithmetic variables from the remaining variables corresponds to a two by three by four <u>bivariate</u> <u>analysis of variance</u> with second-order interactions — this form of prediction occurs at time points when the first two plausibility values are 0 and all other plausibility values are 1.

- Predicting the binary variable from all other variables corresponds to a two-group non-linear discriminant analysis — this form of prediction occurs at time points when the binary plausibility value is 0 and all other plausibility values are 1.
- Predicting the first categorical variable from all other variables corresponds to a three-group nonlinear discriminant analysis — this form of prediction occurs at time points when the first categorical variable plausibility value is 0 and all other plausibility values are 1.
- Predicting the binary variable and both categorical variables from the two arithmetic variables corresponds to a 24-group nonlinear discriminant analysis — this form of prediction occurs at time points when the arithmetic variable plausibility values are 1 and all others are 0.
- Predicting one arithmetic variable from the other variables corresponds to a two by three by four analysis of covariance with second-order interactions and (additively) different covariance slopes for different binary and categorical variable values — this form of prediction occurs at time points when the first plausibility value is 0 and all other plausibility values are 1.

In a similar way, the rapid learning system deals with any number of models involving higher-order powers of arithmetic, binary, and categorical variables. Third-order models may include two-way and three-way interactions, as well as second-order and third-order arithmetic variable powers and cross-products; fourth-order models may include two-way through three-way interactions, as well as second-order through fourth-order interactions; and so on.

As a second example involving several binary variables, the rapid learning system may be applied to solve the well-known parity problem [14]. The parity problem involves a number of binary bit variables at each time point. For example, if 4 bit variables exist then an even parity variable depends on the sum of the 4 bit values in the following way. If the sum is an odd number the parity variable is set to 0; otherwise, the parity variable is set to 1. The parity problem is interesting for the following reasons. First, the binary variables as they stand are not individually correlated with the parity variable at all. Also, products among the binary variables up to the third order are not individually correlated with the parity variable at all. However, the parity variable may be perfectly predicted from a linear combination of all products among the binary variables from the first order through the fourth order.

The rapid learning system can solve the parity problem in the case involving 4 bits, by specifying 4 binary variables that represent each of the 4 bit values along with 1 arithmetic variable representing the parity values. The system can learn to predict the outcome in its usual concurrent way, keeping up with measurement vectors arriving at very rapid rates (see section 2.4.1). For parity problems involving large bit sizes, feature sizes would be too large to allow fast operation. The rapid learning system can solve these problems even more quickly, however, by utilizing special power features (see the end of this section).

The rapid learning system routinely computes powers of arithmetic, binary, and categorical features that involve all possible powers and cross-products among all such features, from second-order up to the user-specified highest order $\overline{p}$ ($\overline{p} = 2, 3, \ldots$). Future versions of *Rapid Learner*™ software that allow users to limit powers and cross-products of a certain order may be written, because many corresponding specialized models are useful. For example, the standard analysis of variance model with interactions involves cross-products of binary and categorical variables, but no arithmetic powers and cross-products [12]. Pending the offering of such extended software options, it should be noted that such specialized models may always be implemented by users *outside* the rapid learning system, simply by supplying the required input features directly as if they were measurements instead of the measurements themselves. To give an example of specialized power feature use, two specialized alternatives to the parity problem are described next.

As described earlier in this section, the parity problem requires predicting parity variable values (sums of several bit variable values) as a function of the bit values themselves. If a user wished to solve the parity problem in the 8-bit case, the rapid learning system would automatically do so if $\overline{p} = 8$ were specified. However, all power features of order 8 would be specified for predicting each of the 9 variables involved as eighth-order functions of the others, while only a small fraction of these features would be needed for predicting the parity variable as a function of the 8 bit variables. The user can speed rapid learning operation considerably in this case by supplying only the necessary features as if they were measurements. The necessary features for solving this problem include the 9 variables as they stand, along with the 248 possible cross-products among the binary measurements.

An elegant alternative to the above approach may be based on the fact that the sum of the 8 binary bits, along with its powers up to order 8, can be combined linearly to predict the parity variable precisely. Using this fact, the user can supply only the parity variable and the sum variable as input measurements, along with a feature specification that produces powers of the sum variable up to order 8. The resulting solution is far faster than the above approach for solving the parity problem, and it can be easily extended to parity prediction when hundreds of bits are involved.

Besides solving the parity problem, which may be only of academic interest, a broad variety of specialized compound feature functions involving powers and cross-products of measurements are useful in a variety of applications. In visual pattern recognition, a variety of spatial features such as edges, corners, and other shapes are simply products of pixel values that define lines, corners, and the like. In many other prediction applications involving several arithmetic variables, only a few product features among those up to a specified degree are needed. For example, in process monitoring and control operations involving many temperature, pressure, and level gauges, only a few product measurements involving ambient temperature gauges may be needed to compensate a few specific pressure gauges for ambient temperature effects.

6.6.2 Historical Compound Features

The rapid learning system computes historical binary, and categorical measurement features in a way that naturally extends historical features of arithmetic measurements only (see section 6.3). Just as the system fits polynomial trends to purely arithmetic measurements, the system also fits polynomial trends to binary and binary category features associated with binary and categorical measurements. The system routinely proceeds in this way when historical trends are specified in the presence of binary and categorical measurements.

In addition to routine historical prediction involving arithmetic, binary, and categorical measurements, the rapid learning system may perform specialized historical prediction as well. For example, historical spatial features may be used to monitor and forecast motion of images within a grid, and historical shape features may be used to recognize and project motion of specialized images within a grid. Users can employ the rapid learning system for motion detection by supplying specialized features as if they were measurements, along the lines of the parity problem example (see section 6.6.1). Alternatively, specialized rapid learning systems may be designed for very fast motion detection by using separable Kernels (see section 6.3.2), along with specialized digital or analog Kernel chips (see section 4.1.3).

CONCLUSION

Future Directions

This chapter presents only a few extended measurement features from a practically limitless realm. Along with the very simple mean features that are introduced, many highly specialized additive features may be used, each of which corresponds to a standard linear structural model [4]. Along with the relatively simple power features that are introduced, highly specialized power features may be used, each of which corresponds to a standard prediction model [13]. Along with the simple spatial features that are introduced, combinations of highly specialized shape and motion features may be used, each of which corresponds to a standard image processing model [8]. Similar features may be introduced that apply to any variety of biological, digital, and analog sensory and signal processing systems. All such features are worthy of future study, because use can be made of learning relationships among all of them under rapidly changing conditions.

Summary

This chapter extends rapid learning and prediction methods from the linear realm to a very general realm. The primary basis for this extension is replacing arithmetic measurements with measurement features. Measurement features introduced in this chapter include powers and cross-products of arithmetic measurements, historical trend functions of arithmetic measurements, and additive functions of arithmetic measurements that are organized in space. Other measurement features include binary and categorical measurement functions, which allow arithmetic, binary, and categorical measurements to be predicted from each other. Still other measurement features include compound feature functions, permitting the use of interactions among binary and categorical measurements, spatial product variables for shape detection, and historical spatial trends for motion detection. The extensions introduced in this chapter are a small sample from a practically limitless realm, within which broad use can be made of learning under rapidly changing conditions.

REFERENCES

1. F.M. Lord & M.R. Novick, *Statistical Theories of Mental Test Scores*, Addison-Wesley, Reading, MA, 1968.
2. M.J. Allen & W.M. Yen, *Introduction to Measurement Theory*, Brooks/Cole, Monterey, CA, 1979.
3. J.E. Laughlin, "Comments on estimating coefficients in Linear Models: 'It Don't Make No Nevermind,'" *Psychological Bulletin*, Vol. 85, pp. 247-253, 1978.
4. W.M. Meredith, "Measurement Invariance, Factor Analysis, and Factorial Invariance," *Psychometrika*, Vol 58, pp. 525-543, 1993.
5. N.L. Johnson & S. Kotz, *Distributions in Statistics: Continuous Univariate Distributions —2*, Wiley, New York, 1970.
6. H. Jeffreys, *Theory of Probability*, 3rd Edn., Oxford, London, 1961.
7. R.J. Jannarone, "The ABC of Measurement," Unpublished Technical Report, Machine Cognition Laboratory, University of South Carolina, 1994.
8. C. Mead, *Analog VLSI and Neural Systems*, Addison-Wesley, Reading, MA, 1995.

9. K. Ma, *Statistical Neural Network Models: Theory and Implementation for Visual Pattern Recognition*, Unpublished Doctoral Dissertation, University of South Carolina, 1991.
10. "Rapid Learning Applications in Visual Processing," *RCNS Technical Report Series*, No. APP96-07, Rapid Clip Neural Systems, Inc., Atlanta, GA., 1996
11. J.P. Box, *R. A. Fisher: The Life of a Scientist*, Wiley, New York, 1978.
12. H. Scheffé, *The Analysis of Variance*, Wiley, New York, 1959.
13. J.D. Finn, *A General Model for Multivariate Analysis*, Holt, Rinehart, & Winston, New York, 1974.
14. D.E. Rumelhart, G.E. Hinton, & R.J. Williams, "Learning Internal Representations by Error Propagation," in D.E. Rumelhart & W.L. McClelland (Eds.), *Parallel Distributed Processing, Explorations in the Microstructure of Cognition,* Vol. 1, pp. 318-364, MIT Press, Cambridge, MA, 1986.

Part 4

Operational Details

Part 4 describes rapid learning system information processing operations performed learning. Chapter 7 covers basic monitoring and forecasting operations, while Chapter 8 covers basic control and refinement operations. As in Part 3, a statistical background will ease understanding of the material in Part 4, although statistical expertise is not essential. Exceptions that require statistical expertise are inserted for information processing specialists and clearly labeled as such. Part 4 should give readers a detailed understanding of the basic monitoring, forecasting, control, and refinement operations that the rapid learning system performs.

7

Monitoring and Forecasting Details

INTRODUCTION . . . 214
7.1. MONITORING DETAILS . . . 214
7.1.1. Arithmetic Monitoring Operations . . . 214
7.1.2. Arithmetic Monitoring Descriptive Statistics . . . 216
7.1.3. Binary and Categorical Monitoring Extensions . . . 218
7.1.4. Basic Monitoring Formulas . . . 219
7.2. FORECASTING DETAILS . . . 220
7.2.1. Forecasting Operations . . . 221
7.2.2. Forecasting Function Separability . . . 223
7.2.3. Basic Forecasting Formulas . . . 225
CONCLUSION . . . 227
REFERENCES . . . 227

INTRODUCTION

This chapter describes monitoring and forecasting operations in practical and mathematical terms. Section 7.1 describes monitoring operations for deviant arithmetic measurements (see section 7.1.1), related descriptive statistics (see section 7.1.2), and binary as well as categorical extensions (see section 7.1.3). Section 7.2 describes forecasting operations, including the use of current and recent history observations to predict future operations (see section 7.2.1). Section 7.2 also describes how forecasting operations may be separated into parallel processing operations if necessary, to increase real-time operating speed (see section 7.2.2). Related learning operations are not described in detail in this chapter, because they are covered elsewhere (see chapter 5, chapter 6). As in previous chapters, mathematical details may be skipped without missing essential information (see sections 7.1.4 and 7.2.3).

7.1 MONITORING DETAILS

Monitoring occurs at each time point. Monitoring begins by predicting each current measurement value as a function of recently observed measurements and/or other current measurements. Once each current measurement has been predicted, each predicted value is compared to each observed value to produce a deviance statistic. The deviance statistic is then provided to the user for graphical display (see section 8.1). The deviance statistic may also be provided to the rapid learning system for learning weight control and feature function refinement (see section 4.1).

7.1.1 Arithmetic Monitoring Operations

Returning to the strain gauge monitoring example, suppose that the rapid learning system receives 250 strain gauge measurement values at each time point. Then prior to any such time point, the system will have learned to predict each strain gauge value as a function of all 249 others (see chapter 5, chapter 6). The first monitoring operation that *Rapid Learner*™ software performs is to compute these predicted values, $\hat{y}_1^{\{t'\}}$ through $\hat{y}_{250}^{\{t'\}}$ ($t' = 1, 2, \ldots$). Based on these predicted values, 250 error values are computed of the form, $y_1^{\{t'\}} - \hat{y}_1^{\{t'\}}$ through $y_{250}^{\{t'\}} - \hat{y}_{250}^{\{t'\}}$. These error values are then used to compute 250 deviance statistics of the following form:

$$d_{j'}^{[\mathrm{E}]\{t'\}} = (y_{j'}^{\{t'\}} - \hat{y}_{j'}^{\{t'\}}) / v_{j'}^{[\mathrm{E}]\{t'-1\}1/2}, \; j' = 1, 2, \ldots, 250$$

(see section 5.2). In this way, deviance statistics are rescaled, taking into account the magnitudes of error statistics that have been learned in the past.

Monitoring operations routinely produce monitoring statistics including predicted current values, deviance values, and error variance values for each measurement. These monitoring statistics in turn govern rapid-learning graphical displays. *Rapid Learner*™ graphical display software routinely plots predicted and observed values for each measurement at each time point (see section 3.2). These plots also include upper and lower tolerance band values for each measurement at each time point. Tolerance band values for each measurement have the form,

$$(\underline{j}_{j'}^{\{t'\}}, \bar{j}_{j'}^{\{t'\}}) = (\hat{y}_{j'}^{\{t'\}} - c^{[\mathrm{TB}]\{t'\}} v_{j'}^{[\mathrm{E}]\{t'-1\}1/2}, \hat{y}_{j'}^{\{t'\}} + c^{[\mathrm{TB}]\{t'\}} v_{j'}^{[\mathrm{E}]\{t'-1\}1/2}),$$

where $c^{[\mathrm{TB}]\{t'\}}$ is a positive tolerance band constant ($j' = 1, 2, \ldots, 250$, $t' = 1, 2, \ldots$). This user-supplied constant value determines tolerance band width and has a "confidence level" interpretation (see sections 3.2 and 5.2).

Besides their graphical display use, deviance statistics govern internal *Rapid Learner*™ software control as well as external process control (see section 8.1). A high deviance statistic magnitude may mean one of two things: either an individual measurement has changed or relationships among measurements have changed. In either case, corrective action may be initiated as necessary, either manually or automatically.

If a measurement has changed, perhaps due to an instrument failure, appropriate internal corrective action may include any of the following: setting the plausibility value for that measurement to zero; excluding that measurement from prediction equations for other measurements in the system; and using predicted values of that measurement instead of measured values until the failure has been corrected. On the other hand, if relationships among measurements have changed, removing the measurement from the prediction system may not be appropriate. Instead, changing the learning weight schedule to accelerate new learning would be appropriate. In this way, the newly discovered relationship can be learned and utilized quickly.

If a deviant measurement value has occurred, a variety of appropriate external corrective actions may be taken as well. The first, essential step involves supplying signals from the rapid learning system to process control devices and/or the user, whenever deviant measurement values occur. Once these deviant values have been supplied and noted, a variety of corrective actions may be initiated, ranging from manual warning initiation to fully automatic control (see section 8.1).

Determining whether a deviant value is caused by instrument failure or by a changing relationship may be straightforward or it may be very subtle. In some cases, patterns of deviance statistics may point toward one conclusion or the other. In other cases, auxiliary descriptive statistics may be needed. Many of these are optionally supplied by *Rapid Learner*™ software (see section 7.1.2).

7.1.2 Arithmetic Monitoring Descriptive Statistics

At every time point, *Rapid Learner*™ software determines if each measurement value is deviant, as indicated by its falling outside its tolerance band. If any measurement is deviant and if the user has selected the monitoring descriptive statistics option (see section 3.2), an array of monitoring descriptive statistics is supplied to the user. These statistics range from very specific statistics for each term in each prediction equation to a single, global deviance index.

Suppose the rapid learning system has determined that two measurements j_1 and j_2 are highly correlated with each other, over a period of trials up to the current time point t' $(t' = 1, 2, \ldots)$. If one of the measurements $j_2^{\{t'\}}$ suddenly becomes deviant, perhaps due to a sudden instrument breakdown, *Rapid Learner*™ software will identify not one but ***both*** measurements as deviant. The magnitude of $d_2^{[E]\{t'\}}$ will be high because the deviant observed value $j_2^{\{t'\}}$ will differ greatly from its predicted value $\hat{j}_2^{\{t'\}}$. In addition, however, the magnitude of $d_1^{[E]\{t'\}}$ will be high because its predicted value $\hat{j}_1^{\{t'\}}$ will be distorted by its deviant independent variable value $j_2^{\{t'\}}$. If only deviance statistics were available to users, then, they would be faced with the dilemma of guessing whether $j_1^{\{t'\}}$ or $j_2^{\{t'\}}$ was the deviant culprit.

To aid users facing this kind of dilemma, *Rapid Learner*™ software provides standardized measurement values. For the above case, the two standardized (zero-mean) values that would identify the culprit are

$$z_1^{\{t'\}} = (j_1^{\{t'\}} - \mu_1^{\{t'-1\}})/v_1^{\{t'-1\}1/2}$$

and

$$z_2^{\{t'\}} = (j_2^{\{t'\}} - \mu_2^{\{t'-1\}})/v_2^{\{t'-1\}1/2}.$$

In this particular case, $j_2^{\{t'\}}$ would be properly identified as the culprit, because its standardized value would be larger than the standardized value corresponding to $j_1^{\{t'\}}$.

As a further aid to pinpointing sources of measurement deviance, *Rapid Learner*™ software supplies prediction statistics for each deviant measurement. These prediction statistics are standardized values of each term in each deviant measurement's prediction equation. Each term is the value of each independent variable's standardized value times its corresponding standardized regression weight. In the above case, all such terms contributing to $\hat{j}_2^{\{t'\}}$ would be small. Also, all such terms contributing to $\hat{j}_1^{\{t'\}}$ would be small, except the term involving $z_2^{\{t'\}}$. Insofar as its standardized regression weight is large, this term would also be large, thus further identifying the culprit as $j_2^{\{t'\}}$.

Along with individual monitoring statistics for pinpointing deviant measurements, global deviance statistics are useful for assessing whether or not deviant measurements may be spurious. When many variables are monitored by the rapid learning system, some of them may be expected to be deviant at each time point, even under normal operating conditions. For example, consider the case where 1,000 instrument readings per second are monitored in a chemical process. If deviance statistics are computed at each time point in the usual way (see section 7.1.1), it is not unusual for them to be normally distributed. Under these conditions, a number of the deviance values are likely to fall outside their tolerance band at each time point, even if no unusual activity has occurred. If the tolerance band constant is set at its default value of 2.0, then about 50 out of 1,000 readings are likely to have deviant values at each time point [1].

The standard global deviance statistic provided by the rapid learning system is denoted by $d^{[G]\{t'\}}$ and known as the *Mahalanobis* distance [2]. For the case involving 1,000 measurements and the same measurement features (see chapter 6), the Mahalanobis distance statistic has the form,

$$d^{[G]\{t'\}} = c_{11}^{\{t'-1\}}(j_1^{\{t'\}} - \mu_1^{\{t'-1\}})^2 + c_{22}^{\{t'-1\}}(j_2^{\{t'\}} - \mu_2^{\{t'-1\}})^2 + \cdots + c_{1000,1000}^{\{t'-1\}}(j_{1000}^{\{t'\}} - \mu_{1000}^{\{t'-1\}})^2 +$$

$$c_{21}^{\{t'-1\}}(j_2^{\{t'\}} - \mu_2^{\{t'-1\}})(j_1^{\{t'\}} - \mu_1^{\{t'-1\}}) + \cdots + c_{1000,999}^{\{t'-1\}}(j_{1000}^{\{t'\}} - \mu_{1000}^{\{t'-1\}})(j_{999}^{\{t'\}} - \mu_{999}^{\{t'-1\}}).$$

The constants in this formula for $d^{[G]\{t'\}}$ depend on covariances among measurements (see section 5.3). If the measurements are completely independent of each other, then their the Mahalanobis distance formula will simplify to the following form:

$$d^{[G]\{t'\}} = (j_1^{\{t'\}} - \mu_1^{\{t'-1\}})^2 / v_{11}^{\{t'-1\}} + (j_2^{\{t'\}} - \mu_2^{\{t'-1\}})^2 / v_{22}^{\{t'-1\}} + \cdots +$$

$$(j_{1000}^{\{t'\}} - \mu_{1000}^{\{t'-1\}})^2 / v_{1000,1000}^{\{t'-1\}}.$$

Thus, in the independent case, the global distance measure is simply the sum of the squared standardized measurement values.

If measurements are not independent of each other, the Mahalanobis distance formula adjusts for dependencies in a way that gives a properly modified global measure of deviance. For example, suppose that every one of 1,000 measurements at a given time point in a record is 1.5 standard deviations from its mean. If the 1,000 measurements are mutually independent, the record is very unusual because every one of 1,000 independent measurements is fairly unusual.

Accordingly, the resulting global distance value of 2,250 will identify it as such ($2{,}250 = 1{,}000 \times (1.5)^2$). However, if all 1,000 measurements are highly dependent to the extent that they are linearly redundant (see section 8.2), the global distance measure will be much smaller because of the way its coefficients are computed (see section 7.1.3). The distance measure should be smaller in this case, because insofar as the measurements are completely redundant, all 1,000 deviance values being 1.5 is the same as only one deviance value being 1.5.

Along with its capacity to account for measurement dependencies, the Mahalanobis distance is a useful global deviance measure for another reason. Just as statistical theory provides a confidence level interpretation for deviance values based on the normal distribution (see section 5.1), it provides a similar interpretation to Mahalanobis distance based on the chi-square distribution [2]. As a simpler alternative, *Rapid Learner*™ software standardizes the global distance measure just like it standardizes error measurements relative to previously observed global distance values. The standardized distance measure has the form,

$$z^{[G]\{t'\}} = (d^{[G]\{t'\}} - \mu^{[G]\{t'-1\}})/v^{[G]\{t'-1\}1/2},$$

where the global distance mean $\mu^{[G]\{t'-1\}}$ and the global distance variance $v^{[G]\{t'-1\}}$ have been learned from previous global distance values (see section 6.1.3). Values of the standardized distance measure may thus be interpreted on the same scale as other deviance values provided by the rapid learning system.

7.1.2 Binary and Categorical Monitoring Extensions

The above deviance and distance statistics are designed primarily for arithmetic measurement monitoring (see section 6.1). However, they may also be used for monitoring binary and categorical measurements. For monitoring binary measurements, all of the above specific and global monitoring statistics may be applied directly. Specially designed deviance statistics are necessary for monitoring categorical measurements, because the rapid learning system converts each categorical measurement to two or more corresponding categorical features (see section 6.4). The standard deviance measures for categorical measurements have the form of Mahalanobis distance measures, with one such measure dedicated to each categorical measurement (see section 7.1.4). Each of these measures at each time point is a function of the binary category vector value corresponding to its observed category at that time point. Each measure also involves the binary category mean vector that has been learned up to that time point, as well as covariances among the binary values that have been learned for that category up to that time point. In addition, each measure is standardized, just as the global Mahalanobis distance measure is standardized (see section 7.1.3). In this way, standardized deviance measures for each categorical variable may be interpreted on the same scale as other deviance values provided by the rapid learning system.

Along with these binary and categorical deviance measures, the rapid learning system also includes descriptive statistics for binary and categorical variables. These statistics are in the form of correct classification tables [3]. For binary variables, each such table has two rows and two columns. Each row represents values of the binary measurement that have been observed (0 or 1), and each column represents values that have been predicted. Each entry represents the number of records for which the row value has been observed at the same time the column value has been predicted. Correct classification performance up to any time point can be seen from the table at a glance. If performance has been perfect, for example, the lower left and upper right table values will be 0.

Categorical contingency tables are natural extensions of binary tables, in that each categorical table has as many rows and columns as its categorical variable has categories. Correct classification performance up to any time point can be seen from each such table at a glance, just as in the binary case. In addition, a variety of standard performance statistics can be computed from these tables [3].

Besides frequency tables for binary and categorical variables, the rapid learning system also provides corresponding relative frequency tables. The entries in these tables are computed recursively in a way that depends on learning weights. This method of computation results in tables having entries that are based on learning impact, in precisely the same way that rapid learning means depend on learning impact (see section 7.1.4).

7.1.4 Basic Monitoring Formulas

(This section covers mathematical details that are not essential for practical use of rapid learning methods.)

(The notation in this section uses conventions that are used throughout the book — see the Glossary.) The basic error variance updating formula for each arithmetic or binary measurement feature is

$$v_{m'}^{[E]\{t'\}} = (l^{\{t'\}}e_{m'}^{[E]\{t'\}2} + v_{m'}^{[E]\{t'-1\}})/(1+l^{\{t'\}}),\ m' = 1, \ldots, m^+.$$

The basic upper and lower tolerance band formulas are

$$\widehat{m}_{m'}^{\{t'\}} + c^{[TB]\{t'\}}v_{m'}^{[E]\{t'\}1/2},\ m' = 1, \ldots, m^+$$

and

$$\widehat{m}_{m'}^{[\{t'\}} - c^{[\mathrm{TB}]\{t'\}} v_{m'}^{[\mathrm{E}]\{t'\}1/2}, \ m' = 1, \ldots, m^+,$$

respectively.

The basic standardized deviate formula is

$$z_{(m^+)}^{\{t'\}} = (\boldsymbol{m}_{(m^+)}^{\{t'\}} - \boldsymbol{\mu}_{(m^+)}^{\{t'-1\}}) \widetilde{\mathbf{D}}(\boldsymbol{v}_{(m^+ \times m^+)}^{\{t'\}})^{-1/2},$$

and the basic standardized regression weight formula is

$$\rho_{(m^+ \times m^+)}^{[Z]\{t'\}} = \mathbf{I}_{(m^+ \times m^+)} - \widetilde{\mathbf{D}}(\boldsymbol{v}_{(m^+ \times m^+)}^{\{t'\}})^{1/2} \boldsymbol{v}_{(m^+ \times m^+)}^{\{t'\}-1} \widetilde{\mathbf{D}}(\boldsymbol{v}_{(m^+ \times m^+)}^{\{t'\}})^{-1/2}.$$

Thus, the standardized contribution of the observed feature $m_{m''}^{\{t'\}}$ to the predicted feature $\widehat{m}_{m'}^{\{t'\}}$ is

$$z_{m''}^{\{t'\}} \rho_{m'm''}^{[Z]\{t'\}}, \ m', \ m'' = 1, \ldots m^+.$$

The Mahalanobis distance-based global deviance function has the form

$$d^{[G]\{t'\}} = (\boldsymbol{m}_{(m^+)}^{\{t'\}} - \boldsymbol{\mu}_{(m^+)}^{\{t'-1\}}) \boldsymbol{v}_{(m^+ \times m^+)}^{\{t'\}-1} (\boldsymbol{m}_{(m^+)}^{\{t'\}} - \boldsymbol{\mu}_{(m^+)}^{\{t'-1\}})^{\mathrm{T}},$$

and the Mahalanobis distance-based deviance function for categorical variables has the similar form

$$d_{j'}^{[C]\{t'\}} = (\boldsymbol{m}_{(\bar{c}_{j'}-1)}^{[C]\{t'\}} - \boldsymbol{\mu}_{(\bar{c}_{j'}-1)}^{[C]\{t'-1\}}) \boldsymbol{v}_{(\bar{c}_{j'}-1 \times \bar{c}_{j'}-1)}^{[C]\{t'\}-1} (\boldsymbol{m}_{(\bar{c}_{j'}-1)}^{[C]\{t'\}} - \boldsymbol{\mu}_{(\bar{c}_{j'}-1)}^{[C]\{t'-1\}})^{\mathrm{T}},$$

$(j' = 1, \ldots, j^{[C]+})$. Each binary categorical mean vector is updated recursively in the usual way (see section 5.3.3), with initial values set at

$$\mu_{j'}^{[C]\{t'-1\}} = (1/\bar{c}_{j'}) 1_{(\bar{c}_{j'}-1)}, \ j' = 1, \ldots j^{[C]+}.$$

Once these deviance measures are obtained, they are converted to standardized statistics in the usual way (see section 7.1).

7.2 FORECASTING DETAILS

Forecasting closely resembles monitoring prediction based on recent history features (see section 6.3), in that each forecast measurement value is predicted as a function of polynomial coefficients, fit to current and past measurement values. Forecasting differs, however, because it requires predicting future measurements instead of current measurements. Therefore, unlike monitoring prediction, the independent variables for forecasting prediction must always be observed before the dependent variables being forecast. Also, unlike monitoring prediction and learning operations, which can occur at the same time point for a given set if independent variables, forecast learning operations require waiting until the future measurement values being predicted have actually been observed.

This section describes forecasting operations that are distinct from monitoring operations because of this future prediction distinction. These distinct operations are first described in operational terms (see section 7.2.1). An interesting special consequence is then described — the capacity of multiple forecasting operations to be separated into simultaneous parallel computing operations (see section 7.2.2). Finally, forecasting operations are described in technical detail (see section 7.2.3).

7.2.1 Forecasting Operations

Suppose that commodities forecasting is being performed using the rapid learning system (see section 1.2.1), to predict future trends for each of 500 commodity prices. Suppose further that the rapid learning system is forecasting each commodity price 1, 3, 5, and 7 minutes into the future. In addition, suppose that each such forecast value for each commodity is being predicted as a function of current values for all commodities, along with their values 2, 4, 6, 8, 10, 12, and 14 minutes into the past. Then at each time point four future values must be computed for each of 500 commodity variables in this example. The result is a total of 2,000 prediction equations that must be computed at each time point.

On the independent variable side of these 2,000 equations, eight current and recent values are involved for each of the 500 commodity variables. The result is 4,000 independent measurement variable values. However, forecasting by the rapid learning system routinely uses historical trend feature values rather than historical measurement values themselves as independent variables. Suppose finally that second-degree trends are used to forecast these 500 commodity values at each time point. Then for each of the 500 commodity variables three coefficients must be computed (constant, linear, and quadratic — see section 6.3). The overall number of independent variables for this example is 1,500 trend features, with the same features being used for each prediction equation.

For this example, then, at each time point 2,000 equations must be solved, each involving 1,500 independent variables and one dependent variable. In addition, three trend values must be computed from seven historical values, for each of the 500 commodity variables. Furthermore, at each time point learning must be updated, including 3,500 means (2,000 dependent variable means + 1,500 independent variable means), three million regression weights (2,000 equations × 1,500 regression weights per equation), and several other parameters (see section

5.3.3). *Rapid Learner*™ software performs all these computations in about five seconds per time point (see section 2.4.1), and rapid learning hardware will perform them all considerably faster (see section 7.2.2). Describing precisely how the rapid learning system performs these operations so quickly is beyond the scope of this book. However, sufficient detail is provided below for readers to perform precisely the same operations manually.

In terms of *Rapid Learner*™ software options and rapid learning notation, this example involves the following options and parameters (the notation follows sections 5.1.5, 5.3.3, and 7.2.3):

- The number of measurements at each time point is $j^{+} = 500$.
- The number of forecasting steps forward in time is $f^{[\text{STEPS}]} = 4$.
- The forecast step size is $f^{[\text{STEP}]} = 2$.
- The number of historical steps backward in time is $h^{[\text{STEPS}]} = 8$.
- The historical step size is $h^{[\text{STEP}]} = 2$.
- The historical trend degree is $\bar{p} = 2$.

The rapid learning system thus performs the following operations at each time point t' ($t' = 14, 15, \ldots$ — see section 4.1.1):

- For each of the commodity variables, it computes three trend coefficient values, ($f_{j'1}^{[\text{T}]\{t'\}}$, $f_{j'2}^{[\text{T}]\{t'\}}$, $f_{j'3}^{[\text{T}]\{t'\}}$, $j' = 1, 2, \ldots, 500$).
- For each of the commodity variables, it forecasts each of four future values ($\widehat{j}_{j'}^{\{t'+1\}}$, $\widehat{j}_{j'}^{\{t'+3\}}$, $\widehat{j}_{j'}^{\{t'+5\}}$, $\widehat{j}_{j'}^{\{t'+7\}}$, $j' = 1, 2, \ldots, 500$), with each value computed as a function of each of the 1,500 trend coefficients.
- It uses these values, along with learned error variances for each of them, to construct forecast *telescopes* (see section 1.2).
- It stores the three trend values associated with each of the 500 commodity variables for future learning use, once the forecast values actually become observed.
- It updates learned parameters for forecasting each commodity value one point into the future, based on currently observed $j_{j'}^{\{t'\}}$ values along with trend values that were computed and saved at time point $t'-1$ ($j' = 1, 2, \ldots, 500$).

  It updates learned parameters for forecasting each commodity value three points into the future by using currently observed $j_{j'}^{\{t'\}}$ values along with trend values that were computed and saved at time point $t'-3$ ($j' = 1, 2, \ldots, 500$).

- It updates learned parameters for forecasting each commodity value five points into the future by using currently observed $j_{j'}^{\{t'\}}$ values along with trend values that were computed and saved at time point $t'-5$ ($j' = 1, 2, \ldots, 500$).
- It updates learned parameters for forecasting each commodity value seven points into the future by using currently observed $j_{j'}^{\{t'\}}$ along with trend values that were computed and saved at time point $t'-7$ ($j' = 1, 2, \ldots, 500$).

It should be noted that each of the required 2,000 forecasting prediction equations uses the same independent variables, but each requires its own distinct set of regression weights. Functionally, predicting and learning operations for each such equation are precisely the same as other similar operations performed by the rapid learning system (see chapter 5 and chapter 6). The only difference is that forecast learning based on a given dependent variable value must occur after the value has been predicted, because the forecast value has not yet been observed at the time it has been predicted.

As with other examples described in this book (see chapter 1), using alternative methods and software to carry out the above operations would require many hours of computer time and manual effort *at every time point*. Instead, *Rapid Learner*™ software performs the same operations in a few seconds by using a variety of short-cuts.

7.2.2 Forecasting Function Separability

Figure 7.2.2 shows parallel prediction and learning structure for forecasting a commodity price (see section 1.2). The forecasting parameters for the example are the same as those given in the previous forecasting example (see section 7.2.1). The vertical axis is the commodity price and the horizontal axis is time in minutes during a trading morning. The figure shows price values starting at 10:49 A.M. and continuing to the current time point, 11:09 A.M. Surrounding the plot are several rapid learning components, which are described later in this section. The figure also shows lines representing component output connections, along with some of their input connections.

As Figure 7.2.2 shows, one monitoring component and four forecasting components are needed in this case (in keeping with $f^{[\mathrm{STEPS}]} = 4$ in the section 7.2.1 example), in order to predict four future time points two minutes apart (in keeping with $f^{[\mathrm{STEP}]} = 2$ in the example). The monitoring component predicts the

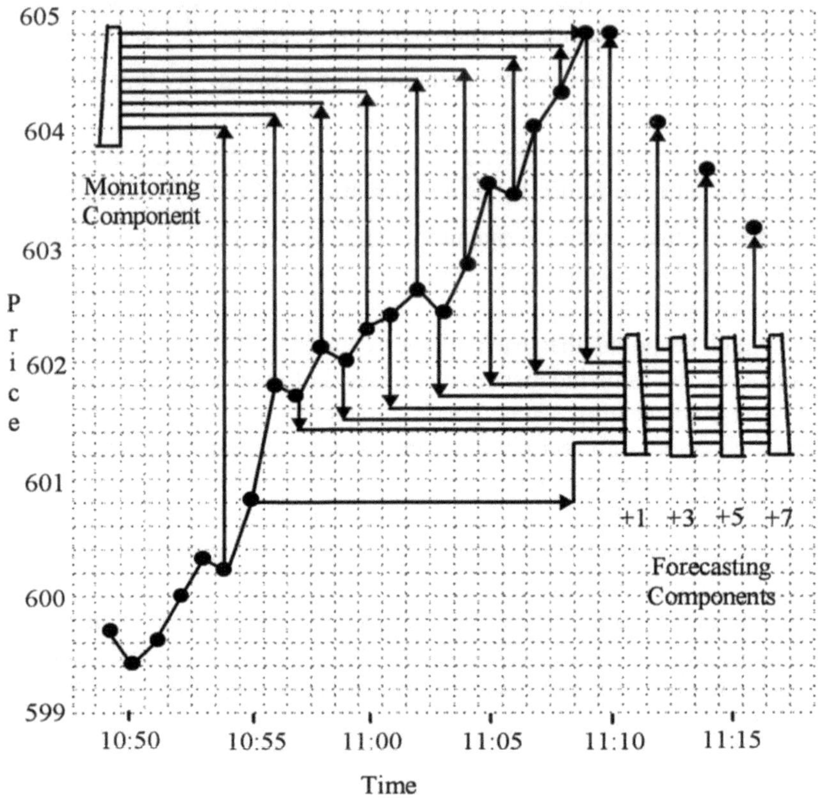

FIGURE 7.2.2 Separable Prediction and Learning Components
(Courtesy of Rapid Clip Neural Systems, Inc.)

price value at the current time point, as indicated by the connection pointing from the component to the 11:09 point. Each forecasting component predicts one of four future price values, as indicated by the connection from the forecasting component labeled "+1" to its corresponding time point, 11:10, and so on.

Figure 7.2.2 also shows input values from the plot that are needed to predict output values on the plot. In keeping with the historical parameters from the forecasting example ($h^{[STEPS]} = 8$ and $h^{[STEP]} = 2$), input values to the monitoring prediction component include measurement values one minute ago, three minutes ago, and so on up to 15 minutes ago. Likewise, input values to the forecasting prediction component include the current measurement value along with values two minutes ago, four minutes ago, and so on to up to 14 minutes ago.

Input values from other measurements that are needed to predict output values in Figure 7.2.2 are not shown. Assuming that the measurement in the figure is one of 500 that are being forecast (see section 7.2.1), each of the input lines shown in the figure represents only one of 500 such lines. For example, the input to the four forecasting components at 11:09 is one of 500 inputs to them at that time; the input to the forecasting component at 10:55 is one of 500 inputs it at that time; and so on. Likewise, the components shown in Figure 7.2.2 are only one set among 500. Each set is needed to monitor and forecast each of the 500 commodity prices, and each requires the same number of inputs as the set in Figure 7.2.2.

The same components, input connection lines, and output connection lines are required at every time point during real-time operation. However, the input and output connection points change from time to time. For example, if the current time point were 11:04 instead of 11:09, the same monitoring Kernel and forecasting Kernels as shown in Figure 7.2.2 would be employed. However, each connection would be made to a measurement occurring 5 minutes earlier than shown.

Each of the components shown in Figure 7.2.2 performs standard rapid learning operations. These include a standard Transducer feature computing operation (see sections 4.1 and 6.3), along with standard Kernel predicting and learning operations (see sections 4.1 and 4.5). Thus, each component includes two modules within it: a Transducer module for computing recent history trend features, and a Kernel module for predicting and learning one output value from many input trend values.

Two aspects of the component structure shown in Figure 7.2.2 make sequential implementation simple and massively parallel implementation feasible. First, all 500 of the monitoring components are identical to each other and all 2,000 of the forecasting components are also identical to each other. As a result, sequential implementation simply involves performing precisely the same Transducer and Kernel operations repeatedly, while changing only the inputs and outputs between repetitions. Second, all 2,500 of these component operations may be performed completely independently and non-iteratively. As a result, all component operations may be performed simultaneously and in parallel. The overall consequences are high software speed — all component operations may be performed at each time point in a few seconds — and very high hardware speed — 2,500 components made up of parallel digital modules (see sections 4.1.2 and 4.1.3) could perform all required component operations in about 100 microseconds, and parallel analog modules could perform the same operations in about 100 nanoseconds (see section 2.4).

7.2.3 Basic Forecasting Formulas

(This section covers mathematical details that are not essential for practical use of rapid learning methods.)

(The notation in this section uses conventions that are used throughout the book — see the Glossary.) Consider the case where each of j^+ measurements must be forecast $f^{[\mathrm{STEPS}]}$ time points into the future, the first of which is one point into the future and the rest of which are $f^{[\mathrm{STEP}]}$ time points apart (j^+, $f^{[\mathrm{STEPS}]}$, $f^{[\mathrm{STEP}]} = 1, \ldots$). Suppose that each of these are forecast as a function of a polynomial of degree $\bar{p}$, fit to $h^{[\mathrm{STEPS}]}$ current and historical values, each of which is separated by $h^{[\mathrm{STEP}]}$ time points ($h^{[\mathrm{STEPS}]}$, $h^{[\mathrm{STEP}]} = 1, \ldots$; $\bar{p} = h^{[\mathrm{STEPS}]}-1$, $h^{[\mathrm{STEPS}]}, \ldots$). Forecasting operations at each time point t' include the forecast prediction and forecast learning ($t' = \max\{f^{[\mathrm{STEP}]}(f^{[\mathrm{STEPS}]}-1)+1, \{h^{[\mathrm{STEP}]}(h^{[\mathrm{STEPS}]}-1)+1\}, \ldots$). Forecast prediction involves computing

$$\hat{j}^{[\mathrm{F}]\{t'\}}_{j'f^{[\mathrm{STEPS}]'}} = \mu^{[\mathrm{FD}]\{t'-1\}}_{j'f^{[\mathrm{STEPS}]'}} + \sum_{j'=1}^{j^+}\sum_{p'=1}^{\bar{p}}(m^{[\mathrm{HF}]\{t'\}}_{j'p'} - \mu^{[\mathrm{FD}]\{t'-1\}}_{j'p'})\rho^{[\mathrm{FD}]\{t'-1\}}_{j'p'}$$

($j' = 1, \ldots, j^+$; $f^{[\mathrm{STEPS}]'} = 1, \ldots f^{[\mathrm{STEPS}]}$), based on historical features of the form,

$$\boldsymbol{m}^{[\mathrm{HF}]\{t'\}}_{(\bar{p}+1)} = \boldsymbol{j}^{[\mathrm{HF}]\{t'\}}\boldsymbol{p}(\boldsymbol{p}^{\mathrm{T}}\boldsymbol{p})^{-1},$$

where

$$\boldsymbol{j}^{[\mathrm{HF}]\{t'\}}_{(h^{[\mathrm{STEPS}]})} = (j^{\{t'-h^{[\mathrm{STEP}]}(h^{[\mathrm{STEPS}]}-1)\}}, \ldots, j^{\{t'-h^{[\mathrm{STEP}]}\}}, j^{\{t'\}})$$

and

$$\boldsymbol{p}_{(h^{[\mathrm{STEPS}]}\times\bar{p}+1)} = \begin{bmatrix} 1 & 1 & \cdots & 1 \\ 1 & 2 & \cdots & 2^{\bar{p}} \\ \vdots & \vdots & \ddots & \vdots \\ 1 & h^{[\mathrm{STEPS}]} & \cdots & h^{[\mathrm{STEPS}]\bar{p}} \end{bmatrix}$$

(see section 6.3.2).

Forecast learning involves updating all means and regression weights in the usual way by using the current learning weight $l^{\{t'\}}$ (see section 5.3). The same historical values that were independent variables for forecasting each of the $\hat{j}^{[\mathrm{F}]\{t'\}}_{j'f^{[\mathrm{STEPS}]'}}$ values in prior trials are used for updating mean and regression weight parameter learning during the current trial.

CONCLUSION

Future Directions

The monitoring and forecasting operations introduced in this chapter may be extended in several ways to increase user friendliness. As currently implemented in *Rapid Learner*™ software, both operations are exhaustive, in that all variables in the system are monitored and forecast as a matter of course. In practice, monitoring or forecasting may be needed for only a few variables, in which case selecting only those variables may speed software operation considerably. Also, forecasting and monitoring specifications are quite rigid as they stand. For example, exactly the same trends are fit to all independent variables in the same way. In practice, correlated trends for some independent variables may have different lag times than correlated trends for others. Giving the user flexibility to specify different trend parameters for different independent variables will thus be useful. At a more basic level, only a few descriptive statistics are currently available out of the many descriptive statistics that are used in other systems. Expanding *Rapid Learner*™ descriptive statistic options in a way that preserves concurrent learning speed is another useful direction for future work.

Summary

This chapter describes monitoring and forecasting operations. Monitoring operations include identifying deviant measurements by comparing observed measurement values to predicted measurement values. Monitoring operations include standardizing deviance statistics to adjust for previously learned deviance values. Monitoring operations also include providing a variety of descriptive statistics that can pinpoint underlying causes for discrepancies between observed and predicted values. Forecasting operations involve predicting future measurements at each time point. Forecasting operations also involve using current time points to update learning. Forecasting operations are potentially much more time-consuming than monitoring operations because separate prediction equations must be used and learned to forecast each time point for each measurement. However, forecast prediction and learning are also separable into component operations. As a result, forecasting may be implemented using efficient software programs, as well as highly efficient, massively parallel hardware.

REFERENCES

1. R.G. Miller, *Simultaneous Statistical Inference*, McGraw-Hill, New York, 1966.

2. C.R. Rao, *Linear Statistical Inference and Its Applications*, 2nd Edn., Wiley, New York, 1973.
3. R.J. Jannarone, C.A. Macera, & C.Z. Garrison, "Evaluating Interrater Agreement Through 'Case-Control' Sampling," *Biometrics*, June, 1987.

8

Control and Refinement Details

INTRODUCTION 230

8.1. CONTROL DETAILS 230

8.1.1. Open-Loop Control 230

8.1.2. Open-Loop Versus Closed-Loop Control 233

8.1.3. Closed-Loop Control 234

8.2. REFINEMENT DETAILS 237

8.2.1. Redundant Feature Clustering 238

8.2.2. Unnecessary Feature Removal 239

8.2.3. Feature Set Expansion 240

8.2.4. Solving a Certain Numerical Problem 242

8.2.5. Basic Refinement Formulas 242

CONCLUSION 246

REFERENCES 247

INTRODUCTION

Chapter 8 describes external control as well as internal refinement operations performed by the rapid learning system. To implement automatic control (see section 8.1), the rapid learning system supplies control signals to processes outside the rapid learning system. To implement automatic refinement, the system uses monitoring and refinement statistics to control feature functions inside the rapid learning system (see section 8.2). In either case, automatic control and refinement add complexity to rapid learning system operation because the system affects not only the way measurements are being predicted but also the way they are being produced (see section 8.1.2).

8.1 CONTROL DETAILS

This section describes two types of control: the open-loop type, where the control signal itself does not affect independent variables that govern it; and the closed-loop type, where the control signal does affect them. (In order to place them in rapid learning system context, open-loop and closed-loop control are described in different terms here than in the classical control literature — see [1] .) Both types are illustrated with an example: open-loop control for missile tracking (see section 8.1.1) and closed-loop control for maintaining level in a tank (see section 8.1.2). Some important distinctions between the two types are described as well (see section 8.1.3).

8.1.1 Open-Loop Control

Returning to the missile tracking example (see section 1.3.1), suppose that the rapid learning system is forecasting each of three missile coordinates one time frame into the future (i.e., $f^{[STEPS]} = 1$ — see section 7.2.1). Suppose further that the system forecasts each coordinate as a function of seven other measurements (i.e., $j^{+} = 10$), along with a recent history of measurements going back five consecutive points in time (i.e., $h^{[STEPS]} = 5$, $h^{[STEP]} = 1$). In addition, suppose that each set of five consecutive measurements is converted to three second-degree trend features (i.e., $\overline{p} = 2$). Given this configuration, the number of features involved for predicting each future coordinate value is 31 (31 = 10 measurements × 3 trend features per measurement + 1 forecast coordinate).

Control response rates depend on several factors. When 31 features per prediction are involved, *Rapid Learner*™ software can forecast and learn within two milliseconds per frame (see section 2.4.1). The software can operate about ten times faster using a fast work-station, and much faster using special-purpose, rapid learning hardware (see section 2.4). Camera control also requires other operations, the performance of which will decrease overall response rates. These operations may include measuring and receiving independent variable values, transmitting predicted values to control systems, converting predicted values to control signals, and using the signals to control positioning devices.

Control accuracy depends on prediction and learning speed, especially under conditions when prediction parameters are changing. In many related applications, control accuracy tends to decrease as the time span increases between learning and prediction time frames. During missile tracking, for example, a missile flight path may be altered by many influences, ranging from varying atmospheric conditions to varying missile mechanics. Such *random walk* influences [2], caused by changing prediction parameters, can be minimized by adjusting learned prediction parameters as quickly as possible.

Control accuracy also depends on forecasting time specifications. For some open-loop control applications, predicting only the next-position coordinates may be sufficient for precise control. For others, additional predictions and prediction statistics may be useful as well. For example, a camera positioning device may be better equipped to keep on track if it is supplied at each time point with three forecast telescopes (see section 7.2.1), rather than only three predicted coordinates for the next time frame. The future trajectories in these telescopes along with their tolerance bands may help keep the camera on track during a complex maneuver. This additional information may be obtained just as quickly by the rapid learning system, since the system is able to learn and predict the additional information separately and at the same time (see section 7.2.2).

Control accuracy depends on forecasting prediction specifications as well. If many measurements are potentially correlated with outcome, the best way to minimize tolerance bandwidth and maximize control accuracy is to include all such measurements as prediction variables. For example, suppose that a missile is being tracked by many radar installations and measured by a variety of other sensors, each of which is subject to *random measurement error* [2]. In that case, including all such sensors as prediction variables will increase prediction accuracy, which in turn will increase control accuracy.

To demonstrate the key role of rapid learning on response accuracy, a simulation study was performed [2]. In the study, missile motion including a set of sudden maneuvers in three dimensions was simulated. The study included a variety of trials involving different levels of random walk error and random measurement error (see section 8.1.4). At all levels of random walk error, camera tracking accuracy increased with rapid learning and prediction response time.

Before implementing new technology such as the rapid learning system for control, process engineers must be convinced that a control system will work in practice, especially when expensive and potentially dangerous equipment is involved. Among other new control systems, the rapid learning system has a key advantage: it may be installed to *monitor* control activity in a harmless way, prior to being deployed as a control device. In that way users can take time to assess rapid learning control signal accuracy and adaptivity, prior to replacing or augmenting an existing control system. Meanwhile, the system can effectively monitor for unusual behavior of existing control equipment.

As an example of rapid learning monitoring use in an open loop control setting, suppose that a camera must be kept on an object in a continuous process, much the same way as in the above missile tracking example — by using sensors that continuously measure the position of the object. Whenever the object speeds up, the camera must also speed up. As a result, if the camera is sluggish during speed-up, it will focus behind the object. Likewise, if the camera is sluggish during slow-down, it will focus in front of the object. Similar problems may occur if the camera anticipates speed-up or slow-down too quickly. The rapid learning system can monitor tracking accuracy effectively, simply by computing an accuracy statistic at each time point and assessing whether or not its values are unusual from time to time. For example, a simple accuracy measure would be the average squared deviation between actual position and the camera focal point over the last 10 time frames. More sophisticated monitoring may also be obtained simply by computing several deviance statistics, each of which measures unusual activity in different time frames.

Along with identifying unusual control activity, initial monitoring use of the rapid learning system has an added benefit: its potential for improving control can be assessed prior to deployment. Many existing control systems, which are not at all adaptive, work quite well, especially in applications where relationships among measurements do not change over time. Other non-adaptive control systems work poorly, because they are used under changing conditions that they cannot deal with. A key indicator of whether or not existing control systems are sufficiently adaptive is how much control performance varies over changing conditions — an indicator that the rapid learning monitoring system is ideally suited to assess. If assessment results indicate that the existing control system is not sufficiently adaptive, then these same results will clearly indicate that a rapid learning system can add control precision. On the other hand, if monitoring results show that the existing control system works well, then the rapid learning system should only be used more passively as a monitoring device.

Two important benefits make using the rapid learning system for monitoring prior to using it for control a good idea:

- Assessing existing control system performance.
- Assessing rapid learning system stability.
- Assessing rapid learning system added value.

8.1.2 Open-Loop Versus Closed-Loop Control

Open-loop control is relatively simple because the measurements that are used for open-loop control are not affected by the control process. For example, keeping a camera on a missile does not effect missile navigation or missile measurement correlations. By contrast, closed-loop control is much more complicated because control signals themselves affect the measurement process (providing so-called *feedback* to close the so-called *control loop* — see [1]). For example, using missile position coordinates to keep a camera on a missile is a simple open-loop process, while keeping a homing device on a missile that is evading the device is a more complicated closed-loop process. By taking evasive action once it senses the homing device, the missile is changing parameters that are being measured by the device. (Open-loop versus closed-loop control considerations have interesting counterparts ranging from relativity concerns in quantum physics to local dependence concerns in psychometrics — see [3]).

Most control applications are closed-loop in nature. The following examples are representative. In an automotive cruise control system, accelerator and transmission control has a direct effect on speed sensors that provide input measurements to the system. In a tank level control system, input feed pump flow control has a direct effect on tank level inputs to the system; in a room temperature control system, heating, air-conditioning, and air flow control has a direct effect on room temperature inputs. In a muscle control system, muscle motion being controlled has a direct effect on visual and tactile input neuron signals.

Closed-loop control is potentially more complicated than open-loop control, because the effects of control actions on input measurements must be recognized and utilized. When a missile begins an excursion without reacting to a homing device, a certain model must be learned for camera control. A different model must be learned for homing device control, however, when a missile is actively evading the device. Although the two models are different, the rapid learning system may learn effectively in either case. Furthermore, the same rapid learning speed and precision advantages of the system apply in the closed-loop case as in the open-loop case (see section 8.1.3).

Using the rapid learning system for control monitoring prior to installing the system for control has the same advantages for closed-loop applications as for open-loop applications. In both cases, existing system adequacy can be verified, inadequacies can be identified, and added value of a rapid learning system can be assessed. In closed-loop applications, however, preliminary monitoring can only assess rapid learning control utility up to a point. Since closed-loop control affects measurement values and correlations, the only way that rapid learning control can be assessed definitively in closed-loop settings is to actually use the system for control.

Engineers may be naturally reluctant to use the novel rapid learning system in place of more familiar systems to control critical processes. Trying out the system in experiments or simulations may be a useful alternative in some applications, but not in others. Taking the time to try out a new system in an operating process can be expensive and potentially dangerous. The alternative approach of simulating the effects of an actual closed-loop control system on measurement values is also difficult, as well as being less definitive than fully operational testing.

For these reasons, preliminary monitoring is especially important in closed-loop applications. In addition, a cautious program of gradual control assessment, analysis, and installation may also be important. A program of this type can begin with monitoring existing control operations and identifying conditions that produce poor control. The next step can include installing the rapid learning system as if it is generating closed-loop control signals (see section 8.1.3), but continuing to operate with the existing system. In this way, discrepancies between the two signals can be identified and analyzed. At the same time, utility of the rapid learning system for control can be partially assessed, without affecting the process being controlled. The next step can include switching from the existing control system to the rapid learning system or augmenting existing signals with rapid learning signals, under relatively safe and carefully monitored operating conditions. The final step prior to complete installation can include exhaustive testing of the rapid learning system, under relatively safe and carefully monitored operating conditions.

As in the open-loop case, it is important to note that while the rapid learning system is being assessed as a potentially valuable control device, it will also be performing a very valuable operation at the outset: monitoring existing control equipment to assess its effectiveness and give early warning of impending problems. Many expensive and dangerous problems due to faulty control equipment can be avoided before they occur, by effective monitoring from the rapid learning system. Furthermore, the entire array of rapid learning advantages apply to control monitoring as well — especially the capacity to monitor with high sensitivity under rapidly changing plant conditions.

8.1.3 Closed-Loop Control

A variety of closed-loop control systems are used in practice, ranging from traditional PID systems to emerging adaptive systems [1]. These systems are applied to many types of control, resembling the tank level control example that is described below (see section 1.3). Most control systems share basic features with PID controllers, including the following provisions for tank control (see Figure 1.3.2):

- A basic control signal that opens an inlet valve if level drops below a set-point and closes if level raises above the set-point (the P for *proportional* in PID).
- A *dampening* signal that suppresses the signal as necessary to prevent overshoot (the D for *differential* in PID).
- Other signals to prevent valve oscillations and tank level drift (the I for *integral* in PID).

Typically, PID controllers adjust control signals in anticipation of their effects on outcome variables, without predicting or forecasting outcome values explicitly.

In contrast to conventional control signals, rapid learning control signals depend directly on predicted outcome values. Continuing with the tank level control example, suppose that a rapid learning prediction model is established, based on a feature vector $\boldsymbol{m}^{\{t'\}}$, made up of the following values at each time point t' ($t' = 1, 2, \ldots$):

- A tank level discrepancy feature value $m^{[\mathrm{LD}]\{t'\}}$, computed as the difference between observed and desired tank level at each time point.
- A control valve position value $m^{[\mathrm{CVP}]\{t'\}}$.
- A vector of other independent variable feature values $\boldsymbol{m}^{[\mathrm{I}]\{t'\}}$, which may include tank temperature values, downstream valve position values, and other current as well as recent values for measurements correlated with tank level.

Suppose further that a rapid learning forecast model has been established (see section 7.2), which produces forecast tank level discrepancy values one time point into the future, $\hat{m}^{[\mathrm{LD}]\{t'+1\}}$, at each time point. Finally, suppose that a valve control signal $s^{[\mathrm{VC}]\{t'\}}$ is generated at each time point with one of three values and outcomes: a value of 0 produces no valve change, a value of +1 opens the valve at a constant rate, and a value of −1 closes the valve at a constant rate.

A simple rapid learning tank control system may operate based on the above model as follows: at each time point t' ($t' = 1, 2, \ldots$), the forecast tank level discrepancy $\hat{m}^{[\mathrm{LD}]\{t'+1\}}$ is first computed. Based on its computed value, one of the following valve control signals is generated:

- If the forecast value is positive, indicating that the control valve should be closed, a valve closing signal is generated by setting $s^{[\mathrm{VC}]\{t'\}} = -1$.

- If the forecast value is negative, indicating that the control valve should be opened, a valve opening signal is generated by setting $s^{[VC]\{t'\}} = +1$.
- If the forecast value is at or near 0, indicating that the control valve need not be moved, an appropriate signal is generated by setting $s^{[VC]\{t'\}} = 0$.

In order for this simple control system to operate effectively, several key requirements must be met. First, the valve positioning system must be sufficiently responsive to keep up with any tank level change. Otherwise, tank level may become too high or too low during changing conditions. Second, the rapid learning system must sample measurements frequently. Otherwise, valve positions may continue to change after it becomes unnecessary to do so, causing the control system to over-react. Third, the feature set used by the rapid learning system must be sufficiently large to predict next time-frame position accurately. These three requirements should be already met in applications where existing control systems are being replaced by the rapid learning system.

The final requirement prior to accurate rapid learning control is training the system to predict future tank levels accurately. Training may proceed in one of several ways. The smoothest way is to install the system for monitoring prior to giving over valve position control to it. In this way, the system will have time to learn necessary prediction weights without disrupting control operation. A second way is to produce control signals manually for a period of training time, prior to giving over automatic control to the system. A third way, which may be less smooth, is to simply initialize the outcome value at 0 and begin operations with the actual tank level at its preset value, along with other rapid learning parameters initialized in the usual way (see section 5.2.3). In this way, proper control signals will be generated initially because all independent variable contributions to the outcome forecasting equation will be 0. Then, as the system gradually learns how these variables should be used, tank level prediction precision and system control precision will increase accordingly.

It should be noted that all the usual rapid learning system advantages that apply for forecasting, monitoring, and open-loop control also apply in this simple closed-loop control application. As in other applications, learning occurs very rapidly for closed-loop applications. Rapid learning, in turn, allows the system to control effectively after a few trials, control adaptively under rapidly changing conditions, and learn how to control automatically.

It should be further noted that the above rapid learning system is too simple to solve many standard control problems. However, the rapid learning advantage allows the system to solve them in a straightforward way. For example, valves often tend to wear down quickly when they are given control signals that cause them to shift frequently. Such problems can be avoided by introducing change features into the rapid learning system as additional outcome measures, much as they are used by PID controllers. Second, controlled variables such as tank levels often go through excursions when plant conditions suddenly change. Under such conditions, using only next-frame forecast values as outcome variables may not be sufficient to produce control signals that will minimize excursions. Straightforward rapid learning solutions to excursion problems include basing control on outcome several time frames into the future instead of only the next time frame. In similar ways, the key rapid learning advantage of precise, adaptive, and automatic prediction can be tailored to suit a variety of other control needs.

8.2 REFINEMENT DETAILS

This section outlines the basis for rapid learning refinement operations that are described in other sections of this book (see sections 1.4 and 3.5; see also chapter 4). The section begins by describing two ways that the feature set may be reduced: either through redundant feature clustering (see sections 8.2.1, 1.4.2, 3.4, and 6.1) or unnecessary feature removal (see sections 8.2.2 and 1.4.4). The section next describes feature set expansion (see sections 8.2.3 and 1.4.3), which adds new measurement features systematically after other features have been clustered or removed. The section ends with a description of other necessary adjustments associated with refinement (see section 8.2.4) and technical refinement details (see section 8.2.5).

Model refinement marks a distinct point of departure for the rapid learning system, because refinement operations require more time and information processing resources than real-time operations. The distinction between model refinement and real-time learning has some interesting biological counterparts. In evolutionary terms, model refinement represents cerebral cortex operations, which have only matured recently on the evolutionary scale. In cognitive terms, model refinement represents thought while real-time operation represents automatic learning and behavioral activity (see section 4.3). In response time terms, both refinement operations and thought processes take more time than real-time operations. While low-level cognitive activity can occur during routine behavioral operations, higher-level activity requires "stopping to think" and sleeping [4]. In mental activity terms, low-level cognitive activity can be performed by separate processors performing in parallel with real-time operations (see section 4.1), while higher-level activity may demand the use of all available processing resources. In either case, routing rapid learning operations must be performed separately from occasional refinement operations.

The rapid learning system separates routine rapid learning operations from occasional refinement operations by using model refinement modules that are separate from rapid learning modules (see section 4.1). Occasionally, routine rapid learning operations are interrupted briefly, so that currently learned connection weights may be copied from the Learned Parameter Memory to the Manager, where refinement operations occur. After this brief interruption, routine rapid learning operations continue while refinement operations commence. At the end of refinement operations, an appropriate new model is configured while routine rapid learning operations continue. At that point, rapid learning operation is again interrupted and the new model is implemented in place of the old model. The process continues in this way, marked by occasional interruptions and followed by refinement operations.

The refinement process is initiated periodically by the rapid learning system. The initiation frequency can be set by the user, or it can be fixed by default. The default initiation frequency set by *Rapid Learner*™ software is every 100 records (see section 3.5). In this section each such initiation and subsequent set of refinement operations are called a refinement *cycle*. Each cycle may include any or all of the operations described in this section.

8.2.1 Redundant Feature Clustering

Redundant feature clustering is simple. When two measurements have practically the same correlations with all other measurements, the two measurements are treated as equivalent and the rapid learning system replaces them with their mean. Using a mean among equivalent measurement values instead of the values themselves increases speed (see section 2.4.1), decreases required storage (see section 2.4.2), and produces precise predictions [5]. The same rationale and conclusion holds for any *cluster*, made up of several equivalent measurements.

The rapid learning system bases refinement operations on *partial correlations* among measurement features [5]. Every pair of measurement features has a partial correlation coefficient, which is updated at every time point by the rapid learning system. Each partial correlation coefficient measures how precisely two features can be predicted from each other, over and above all other features currently included in the system. Partial correlation coefficients may also be interpreted directly in tolerance band reduction terms (see section 8.2.4).

At every time point, the rapid learning system can compute a partial correlation coefficient among any pair of features, just as quickly as it can compute regression weights. For example, the partial correlation $\chi_{12}^{[P]\{t'\}}$ between features 1 and 2 can be computed just as quickly as the regression coefficient for predicting feature 1 from feature 2 at each time t' ($t' = 1, 2, \ldots$). Likewise, partial correlation coefficients between any pair of features are readily available for refinement at any time point.

The rapid learning system identifies redundant features by assessing similarities among their partial correlation coefficients. Suppose that the system is monitoring 200 strain gauges of two different types (see section 6.1.1), with the first 100 being one type and the second 100 being the other type. If each of these 200 gauges is equivalent to other gauges of its type, then every pair of gauges within each type will have similar partial correlation coefficients. The rapid learning system assesses similarity among each pair of features by computing a *redundancy distance* measure between each pair of features, based on their partial correlation coefficients. For the first pair of features, the measure has the form,

$$d_{12}^{[R]\{t'\}} = [(\chi_{13}^{[P]\{t'\}} - \chi_{23}^{[P]\{t'\}})^2 + (\chi_{14}^{[P]\{t'\}} - \chi_{24}^{[P]\{t'\}})^2$$

$$+ \cdots + (\chi_{1,200}^{[P]\{t'\}} - \chi_{2,200}^{[P]\{t'\}})^2]/198;$$

for the next pair of features, the measure has the form,

$$d_{13}^{[R]\{t'\}} = [(\chi_{12}^{[P]\{t'\}} - \chi_{32}^{[P]\{t'\}})^2 + (\chi_{14}^{[P]\{t'\}} - \chi_{34}^{[P]\{t'\}})^2$$

$$+ \cdots + (\chi_{1,200}^{[P]\{t'\}} - \chi_{3,200}^{[P]\{t'\}})^2]/198;$$

and so on.

Once the rapid learning system has computed redundancy distance measures for each pair of features, it compares each measure against a small positive cut-off value. If the distance measure is smaller than the cut-off measure, the pair of features is classified into the same cluster. Otherwise, the pair is not. Once all features have been classified in this way, each cluster is replaced by the mean feature for that cluster (see section 6.1).

This refinement procedure resembles other refinement procedures below in one key way, and it is distinct from real-time rapid learning operations in a second key way. Like other refinement procedures, it is based purely on statistics that can be derived from connection weights (see section 8.2.5), which are stored and maintained in the Learned Parameter Memory (see section 4.1.1). Like other refinement procedures — but in sharp contrast with real-time rapid learning operations — each redundancy operation requires several recursive arithmetic steps. For example, the above clustering operation requires computing a distance measure for *each pair* of measurement features, and each measure is made up of many terms. These computations require much more parallel computing power or conventional computing time than routine prediction and learning operations require.

8.2.2 Unnecessary Feature Removal

Unnecessary feature removal is a relatively simple process that requires two steps:

1. Establishing which essential features must be retained and which potentially unnecessary features may be removed.
2. Removing features from the latter set that add little predictive precision to the former set, one at a time.

The first step may be performed either internally and automatically or externally and manually (with user-supplied plausibility specifications — see section 3.1.6). To perform the first step internally and automatically, *Rapid Learner*™ software classifies features as potentially unnecessary, unless they are either measurements to be predicted themselves or mean features. Optionally, the software also allows users to classify some of the measurements themselves or mean features as potentially unnecessary. For example, users might be supplying some measurements to the system to add precision to the predictability of other measurements, but not to have the measurements monitored or forecast. In that case, users might want to identify these measurements as potentially unnecessary, and *Rapid Learner*™ software and rapid learner gives them the option to do so.

To perform the second step, the rapid learning system relies on partial correlation coefficients (see sections 8.2.1 and 8.2.5). The system first evaluates the partial correlation coefficient between each potential feature for removal (type R) and each feature that must be kept (type K). In the process, the type R feature having the smallest maximum partial correlation coefficient is identified. If that maximum coefficient is larger than a small, pre-established cut-off value, then no features are removed. Otherwise, that feature is removed from the system, the connection weight matrix is adjusted accordingly, and the process is repeated. The process continues until no type R feature remains having all of its partial correlation coefficients below the cut-off value. In order to compute criterion measures for removing unnecessary variables, tolerance band reduction factors are computed as functions of partial correlation coefficient values (see sections 3.5.1 and 8.2.5).

8.2.3 Feature Set Expansion

Once features have been removed by clustering or some other means, storage space or time may be available for new features. The rapid learning system has provisions for introducing such new features manually or automatically. Manual feature set expansion requires interrupting real-time operation, changing feature configurations, and then resuming real-time operation. For example, consider the refinement case study described earlier in this book (see section 1.4), where the feature set was expanded by introducing second-degree cross-products at a certain point (see section 1.4.3). These features could be introduced manually, simply by specifying them explicitly (see section 3.6.2). Alternatively, they could be introduced automatically, by a rapid learning system with the capacity to determine the following:

- The availability of sufficient memory after previous feature removal operations to introduce new second-degree features.
- The availability of sufficient time to both include new second-degree features and keep up with real-time learning and prediction operations.
- The potential for second-degree features to improve predictability.

Once new second-degree features have been introduced, the learning weight schedule should be revised to allow second-degree feature mean values and covariance values to be learned quickly.

In order to make the above determinations automatically, the rapid learning system must be pre-programmed and/or configured accordingly. Acceptable memory sizes must be pre-specified, acceptable record processing rates must be pre-specified, a feature expansion hierarchy must be pre-programmed, and learning schedule modification rules must be pre-specified. Among these decisions, the most difficult and interesting is choosing a feature function hierarchy, that is, the order in which new features should be introduced. The rapid learning system has several expansion feature types to choose from (see chapter 6), including power features of differing degrees, historical trend features of different types, spatial features of different types, and compound features. Each feature type has the potential to improve prediction accuracy better in some applications than the others. Deciding the order in which they should be introduced, however, is very difficult to anticipate for any given application.

Thus, whether feature expansion decisions are manual or automatic, deciding which features to add is not easy. The rapid learning system eases the feature addition process considerably, however, because it has two basic advantages: the capacity to include many features at the same time (see section 2.4.2), and the capacity to reduce the number of features quickly (see sections 8.2.1 and 8.2.2). As a result, problems like pattern recognition that require feature expansion can be solved by an alternating feature expansion and removal process. For example, in one case study the system successfully identified necessary second-degree features because it had the capacity to include all possible second-degree features at once (see section 1.4.1). Furthermore, since the system had the capacity to combine redundant features into clusters and remove unnecessary features for prediction [6], the same process could have been repeated if necessary to identify third-degree features.

In its current form, the rapid learning refinement system has only a small number of automatic feature function expansion options available to users. Establishing, implementing, and automating many more options represents perhaps the most interesting challenge for future rapid learning system development.

8.2.4 Solving a Certain Numerical Problem

Besides feature function reduction and expansion, the rapid learning system makes occasional minor parameter adjustments to avoid unstable learning. Learning can become unstable under one of two conditions: (a) the variance of one or more measurement features becomes close to zero, or (b) two or more measurements become nearly linearly redundant, as indicated by very high connection weight mean diagonal elements (see section 8.2.5). The rapid learning system checks for either of these conditions during each refinement cycle (see section 3.5.1). If either of the conditions is sensed, the system adjusts the variances among the features to remove the problem. The adjustment produces the same effect as if learning were updated at that point due to many measurements having completely independent features and yet small learning weights. (The adjustment is the same as if a block of 2^{m^+} redundant prior feature values is processed and learned, based on a block learning impact value of $1/m^+$ — see sections 3.1.7 and 5.3.1.) The adjustment has a minor effect on the prediction of each feature as a function of all others.

8.2.5 Basic Refinement Formulas

(This section covers mathematical details that are not essential for practical use of rapid learning methods.)

(The notation in this section uses conventions that are used throughout the book — see the Glossary.) As in the case of concurrent learning and prediction (see section 5.3.10), concurrent refinement speed and storage efficiency stem from computing required statistics as simple functions of concurrently updated connection weights. Accordingly, this section formulates the key concurrent refinement statistics as functions of connection weights. Refinement statistics to be covered include squared multiple correlations, partial correlations and connection weights adjusted for variable removal.

Consider the case where the squared multiple correlation (SMC) is required between the last element of $\boldsymbol{m}$ and its regression estimate based on all elements of $\boldsymbol{m}$ but the last. From standard results in multivariate statistical analysis [5], the feature correlation matrix can be computed from the connection weight matrix $\boldsymbol{\omega}$ (i.e. the inverse of the covariance matrix), as follows:

$$\chi^{\{t'\}}_{(m^+\times m^+)} = \widetilde{\mathbf{D}}(\nu^{\{t'\}}_{(j^+\times j^+)})^{1/2}\,\omega^{-1}_{(m^+\times m^+)}\,\widetilde{\mathbf{D}}(\nu^{\{t'\}}_{(j^+\times j^+)})^{1/2}.$$

The required SMC can be computed from the feature correlation matrix as follows: if the correlation matrix is partitioned into

$$\chi_{(m^+ \times m^+)} = \begin{bmatrix} \chi^{[11]}_{(m^+-1 \times m^+-1)} & \chi^{[12]}_{(m^+-1 \times 1)} \\ \\ \chi^{[21]}_{(m^+-1)} & \chi_{m^+ m^+} \end{bmatrix},$$

then the required SMC satisfies

$$\chi^{[m^+|1,\ldots,m^+-1]} = \chi^{[21]}\chi^{[11]-1}\chi^{[12]}.$$

Now since [5]

$$\chi^{-1}_{m^+ m^+} = 1 - \chi^{[21]}\chi^{[11]-1}\chi^{[12]},$$

the required SMC resides on the lower right element of the diagonal multiple correlation matrix,

$$\widetilde{\mathbf{D}}(\chi^{[M]}) = \mathbf{I} - \widetilde{\mathbf{D}}(\chi^{-1})^{-1}.$$

By symmetry, the SMC value for each element in $\boldsymbol{m}$ lies in its corresponding diagonal element of the multiple correlation matrix. Also, since [5]

$$\begin{aligned} \chi^{-1}_{(m^+ \times m^+)} &= \widetilde{\mathbf{D}}(v_{(m^+ \times m^+)})^{1/2}\, v^{-1}_{(m^+ \times m^+)}\, \widetilde{\mathbf{D}}(v_{(m^+ \times m^+)})^{1/2} \\ &= \widetilde{\mathbf{D}}(v_{(m^+ \times m^+)})^{1/2}\, \omega_{(m^+ \times m^+)}\, \widetilde{\mathbf{D}}(v_{(m^+ \times m^+)})^{1/2}, \end{aligned}$$

the multiple correlation matrix can be expressed directly in terms of feature variances and connection weights, both of which are easy to update concurrently.

Turning next to partial correlations, if the feature correlation matrix is partitioned into

$$\chi_{(m^+ \times m^+)} = \begin{bmatrix} \chi^{[AA]}_{(2 \times 2)} & \chi^{[AB]}_{(2 \times m^+-2)} \\ \\ \chi^{[BA]}_{(m^+-2 \times 2)} & \chi^{[BB]}_{(m^+-2 \times m^+-2)} \end{bmatrix},$$

then the partial correlation between the last first elements in $\boldsymbol{m}$, adjusted for all but the first two elements in $\boldsymbol{m}$, is

$$\chi^{[P]}_{2,1} = \gamma^{[1,2|3,\ldots,m^+]}_{2,1}(\gamma^{[1,2|3,\ldots,m^+]}_{1,1}\gamma^{[1,2|3,\ldots,m^+]}_{2,2})^{-1/2},$$

where

$$\gamma_{(2\times 2)}^{[1,2|3,\ldots,m^+]} = \chi_{(2\times 2)}^{[AA]} - \chi_{(2\times m^+-2)}^{[AB]} \chi_{(m^+-2\times m^+-2)}^{[BB]-1} \chi_{(m^+-2\times 2)}^{[BA]}.$$

Also, if the correlation matrix inverse is partitioned into

$$\chi_{(m^+\times m^+)}^{-1} = \begin{bmatrix} [\chi^{-1}]_{(2\times 2)}^{[AA]} & [\chi^{-1}]_{(2\times m^+-2)}^{[AB]} \\ [\chi^{-1}]_{(m^+-2\times m^+-2)}^{[BA]} & [\chi^{-1}]_{(m^+-2\times m^+-2)}^{[BB]} \end{bmatrix},$$

then [5]

$$\gamma_{(2\times 2)}^{[1,2|3,\ldots,m^+]-1} = [\chi^{-1}]_{(2\times 2)}^{[AA]}.$$

Now since [5]

$$\gamma_{(2\times 2)}^{[1,2|3,\ldots,m^+]-1} = \left|\gamma_{(2\times 2)}^{[1,2|3,\ldots,m^+]}\right|^{-1} \begin{bmatrix} \gamma_{2,2}^{[1,2|3,\ldots,m^+]} & -\gamma_{2,1}^{[1,2|3,\ldots,m^+]} \\ -\gamma_{2,1}^{[1,2|3,\ldots,m^+]} & \gamma_{2,2}^{[1,2|3,\ldots,m^+]} \end{bmatrix}$$

and since

$$[\tilde{\mathbf{D}}(\chi_{(m^+\times m^+)}^{-1})^{-1/2} \chi_{(m^+\times m^+)}^{-1} \tilde{\mathbf{D}}(\chi_{(m^+\times m^+)}^{-1})^{-1/2}]_{2,1} = \chi_{2,1}^{-1}(\chi_{1,1}^{-1}\chi_{2,2}^{-1})^{-1/2},$$

as well, the desired partial correlation can be obtained from the correlation matrix inverse by computing

$$\gamma_{2,1}^{[1,2|3,\ldots,m^+]} = -[\tilde{\mathbf{D}}(\chi_{(m^+\times m^+)}^{-1})^{-1/2} \chi_{(m^+\times m^+)}^{-1} \tilde{\mathbf{D}}(\chi_{(m^+\times m^+)}^{-1})^{-1/2}]_{2,1}.$$

By symmetry, the partial correlation among any pair of elements in $\boldsymbol{m}$, adjusted for all other elements in $\boldsymbol{m}$, is the corresponding row and column element for that pair in

$$\chi^{[P]} = 2\mathbf{I} - \tilde{\mathbf{D}}(\chi_{(m^+\times m^+)}^{-1})^{-1/2} \chi_{(m^+\times m^+)}^{-1} \tilde{\mathbf{D}}(\chi_{(m^+\times m^+)}^{-1})^{-1/2}.$$

Rapid learning partial correlation estimates thus have the following functional relationship to concurrent connection weight estimates:

$$\chi^{[P]} = 2\mathbf{I} - \widetilde{\mathbf{D}}(\omega_{(m^+ \times m^+)})^{-1/2} \omega_{(m^+ \times m^+)} \widetilde{\mathbf{D}}(\omega_{(m^+ \times m^+)})^{-1/2}.$$

Partial correlation coefficient values are converted to tolerance band reduction values, which in turn are used as criteria for unnecessary feature removal (see section 3.5.1). The tolerance band reduction conversion function is,

$$\widetilde{f}^{[TBR]}(\chi^{[P]}) = (1 - \chi^{[P]2})^{1/2}.$$

Turning next to adjusted connection weights, suppose that the connection weights among all but the last element in $\boldsymbol{m}$ are required, with each weight adjusted for removal of the last element in $\boldsymbol{m}$. If the connection weight matrix is partitioned into

$$\omega_{(m^+ \times m^+)} = \begin{bmatrix} \omega^{[11]}_{(m^+ - 1 \times m^+ - 1)} & \omega^{[12]}_{(m^+ - 1 \times 1)} \\ \omega^{[21]}_{(m^+ - 1)} & \omega_{m^+ m^+} \end{bmatrix},$$

then, following the notation of Section 5.3.10, the adjusted matrix inverse satisfies [5]

$$\nu^{[11]-1}_{(m^+ - 1 \times m^+ - 1)} = \omega^{[11]}_{(m^+ - 1 \times m^+ - 1)} - \frac{1}{\omega_{m^+ m^+}} \omega^{[12]}_{(m^+ - 1 \times 1)} \omega^{[21]}_{(m^+ - 1)}.$$

Thus connection weights may be adjusted for removal of the last element in $\boldsymbol{m}$ by computing straightforward functions of the full connection weight matrix. Also, connection weights adjusted for removing other elements in $\boldsymbol{m}$ may be obtained as similar functions of the original connection weight matrix. Furthermore, efficient and separable algorithms for updating the connection weight matrix, which are the primary basis for rapid learning system speed, may be tailored to computing adjusted connection weights in this way, with only minor alterations.

Finally, the basic standardized deviate formula is

$$z^{\{t'\}}_{(m^+ \times m^+)} = (\boldsymbol{m}^{\{t'\}}_{(m^+ \times m^+)} - \mu^{\{t'-1\}}_{(m^+ \times m^+)}) \widetilde{\mathbf{D}}(\nu^{\{t'\}}_{(m^+ \times m^+)})^{-1/2},$$

and the basic standardized regression weight formula is

$$\rho^{[Z]\{t'\}}_{(m^+ \times m^+)} = \mathbf{I}_{(m^+ \times m^+)} - \widetilde{\mathbf{D}}(\nu^{\{t'\}}_{(m^+ \times m^+)})^{1/2}\, \omega^{\{t'\}}_{(m^+ \times m^+)}\, \widetilde{\mathbf{D}}(\nu^{\{t'\}}_{(m^+ \times m^+)})^{-1/2}.$$

CONCLUSION

Future Directions

The two operations described in this chapter — highly adaptive control and automatic model refinement — are the most open-ended and important of all rapid learning areas for future research. Automatic control systems affect most human lives in industrial society many times a day. Heating, transportation, telecommunications, military equipment, and financial commerce control systems are among the most important, but many others exist. Highly adaptive systems that can identify developing problems — *and* adjust for changes automatically — have the potential to save down-time costs and prevent dangerous incidents. The rapid learning system is ideally suited for adaptive control, but system usefulness — and especially safety — must be demonstrated convincingly before rapid learning control can be widely accepted. The best way to ensure both usefulness and safety is through a variety of carefully designed future case studies that cover a range of applications.

Control systems are based on established control theory that has its own language and theoretical framework. Delivering rapid learning system control to solve practical problems will require describing the system to control engineers in their own language and evaluating the system within their own established framework. The author welcomes critical suggestions from control experts toward that end.

Fittingly, model refinement is the final topic for future directions in this book. Just as model refinement has marked a basic advance in other neurocomputing systems [7], it marks a point of departure for the rapid learning system. The rapid learning system can rely less heavily than other systems on refinement, because of its large network capacity and its high operating speed. Even so, a far broader range of problems will become solvable by future rapid learning systems as their model refinement capability grows.

Biological evolution closely resembles the rapid learning system in this regard. The biological capacity to monitor, forecast, control, and learn in real-time appeared at an early evolutionary stage, while cognitive processes appeared only recently after advanced species began to think. Similarly, the rapid learning system is able to solve an important but limited class of monitoring, forecasting, control, and learning problems as it stands. Broadening this ability into the cognitive realm will require adding substantial new model refinement capacity to the system, *while preserving the basic capacity to learn in real time*. Preserving this basic requirement, in turn, will ensure consistency with the basic biological capacity to function *and* learn at the same time.

Summary

This chapter describes currently available rapid learning methods for control and refinement. Rapid learning control systems may be classified into open-loop or closed-loop types. Open-loop control systems are relatively simple to test and implement, because the control process itself does not effect system input measurements. Since closed-loop systems affect input measurements, related rapid learning systems must be developed and tested with considerably more care than open-loop systems. In either case, rapid learning methods have the same basic advantage for control that has been underscored throughout this book: the capacity to predict outcome values quickly, precisely, *and* adaptively in real-time.

Rapid learning refinement operations are distinguished by their capacity to function separately from real-time monitoring, forecasting, control, and learning operations. In its present form, rapid learning system refinement is limited to simple clustering and step-wise deletion procedures. Current rapid learning and refinement operations also rely on manual methods for introducing new features. Expanding the refinement realm presents the next challenge for the rapid learning system. Meeting this challenge — while preserving real-time learning — will produce new solutions to the main problem of the information age: "*So much to learn . . . so little time.*"

REFERENCES

1. K.J. Aström & T.J. McAvoy, "Intelligent Control: an Overview and Evaluation," in D.A. White & D.A. Sofge (Eds.), *Handbook of Intelligent Control*, Van Nostrand Reinhold, New York, 1992.
2. "Vehicle Tracking Applications," *RCNS Technical Report Series*, No. APP96-04, Rapid Clip Neural Systems, Inc., Atlanta, GA, 1996.
3. R.J. Jannarone, "Conjunctive Item Response Theory: Cognitive Research Prospects," in M. Wilson (Ed.), *Objective Measurement: Theory into Practice*, Vol.1, Ablex, Norwood, NJ, pp. 211-236, 1992.
4. R.D. Hawkins, T. Abrams, T.J. Carew, & E.R. Kandel, "A Cellular Mechanism of Classical Conditioning in Aplysia: Activity-Dependent Amplification of Presynaptic Facilitation," *Science*, Vol. 219, pp. 400-405, 1983.
5. T.W. Anderson, *An Introduction to Multivariate Statistical Analysis*, 2nd Edn., Wiley, New York, 1984.

6. Y. Hu, *Concurrent Information Processing with pattern Recognition Applications*, Unpublished Doctoral Dissertation, University of South Carolina, 1994.
7. D.E. Rumelhert & W.L. McClelland (Eds.), *Parallel Distributed Processing, Explorations in the Microstructure of cognition*, Vol. 1, MIT Press, Cambridge, MA, 1986.

Glossary

adaptive prediction Prediction that changes either due to changing prediction coefficients that are updated by the rapid learning system and/or changing independent variable values, as established by off-line analysis.

analysis of covariance A statistical analysis method using arithmetic and binary or categorical independent variables, along with an arithmetic dependent variable. A multivariate version uses multiple arithmetic dependent variables.

analysis of variance A statistical analysis method that uses binary or categorical independent variables, along with an arithmetic dependent variable. A multivariate version uses multiple arithmetic dependent variables.

arbitrage A trading method based on identifying discrepancies between actual and predicted price during fast trading activity and placing orders accordingly, sometimes automatically.

arithmetic variables Variables used in linear prediction functions by the rapid learning system.

axon Output appendage to a neuron.

backpropagation learning An iterative process for adjusting connection weights in multi-layer perceptron models. The process uses a training sample, identifies discrepancies between expected and computed outputs, and adjusts connection weights accordingly. For each record, the process adjusts connection weights systematically from output to input.

binary variables Variables that have only two possible values, 0 or 1.

bit An electronic binary storage medium. Computers often organize several bits into larger words.

bivariate prediction Prediction based on two variables, each of which may be a dependent variable.

categorical variables Variables having three or more positive, consecutive integer values from 1 upward.

closed-loop control Prediction involving the use of dependent variable to influence a process in a way that also influences future values of independent variables in the prediction function; as opposed to open-loop control.

cluster A collection of variables that are combined by the rapid learning system to produce an mean feature.

compound features Features that are functions of other features, as opposed to functions of measurements.

connection weights Physical entities that have the same functional role as coefficients in prediction functions.

control The use of prediction function dependent variables to govern a part of a process.

control calibration Establishing or modifying a control prediction function; typically performed off-line.

correlation The relationship between two random variables, usually measured by a coefficient with a magnitude between 0 and 1.

covariance A statistic that reflects the linear predictability of one random variable from another.

dendrite The input appendage to a neuron.

deviance statistics rapid learning system statistics to measure how unusual each measurement is, relative to its predicted value.

differential reinforcement learning A series of events that do not have the same learning impact on biological learning, or a series of records that do not have the same impact on neuro-computing system learning

discriminant analysis A statistical prediction method that utilizes a binary or categorical dependent variable.

distance values Statistics that increase as the difference between two measurements or measurement vectors increases.

error variance A rapid learning system statistic to measure the difference between previously observed and predicted measurement values.

estimation Off-line, sample-based evaluation of statistical prediction function coefficients.

Executive The rapid learning system module that coordinates overall system operation.

feature The basis for rapid learning system modeling and Kernel module operation. The system converts measurement values in each record to nonlinear and/or linear feature values and supplies them to the Kernel.

feature removal A rapid learning system refinement operation, which deletes unnecessary features from the system.

feature set expansion A rapid learning system refinement operation, which adds new features to the system.

forecasting Predicting future measurement values.

forecasting steps Multiple forecasting time points per record.

forecasting step size The number of time points between consecutive forecasting steps.

fuzzy control A process for which a user specifies control operations in terms of fuzzy indices.

fuzzy index A number between 0 and 1 specified by a user to express subjective belief that an attribute applies to a situation, that a control signal should be sent to a device under certain conditions, etc.

general linear models Statistical prediction functions in which dependent variables are computed as linear functions measurements or of measurement features, which may or may not be linear.

hardware Computing equipment.

historical features Features based on current and recent measurement values.

historical steps Multiple recent measurement time points per record, used to compute historical features.

historical step size The number of time points between consecutive historical steps.

image processing Converting visual pattern measurements to features and utilizing the features for information processing.

information processing The use of data by computers for a variety of purposes, especially adaptive monitoring, forecasting, and control.

interaction effects Non-linear functions of categorical measurements.

iterative neuro-computing estimation Multiple, off-line connection weight adjustments, performed by making multiple passes through a training sample, until distance values between observed and predicted output values satisfy a prespecified criterion.

joint access memory Array of memory elements, switches, and buses that plays a central role in Kernel operation.

judgment value A rapid learning system measurement value.

Kernel The rapid learning system module that predicts each feature as a linear function of other features and updates connection weight values in real time.

Learned Parameter Memory The computer storage medium for rapid learning system Kernel statistics used to compute prediction function values.

learning The establishment of behavior patterns from experience in biological systems, or the identification of input-output connection weights from data based on neuro-computing systems.

learning impact The effect of a record on learning, relative to other records that have been previously observed.

learning weight The ratio of current record learning impact value to the sum of all prior learning impact values.

linear prediction The use of a prediction function comprising a weighted sum of independent variable values and a constant.

linear regression A statistical analysis method involving arithmetic variables based on linear prediction.

Mahalanobis distance A global feature deviance measure that adjusts for co-variance among features.

Manager The rapid learning system module that controls feature function refinement and learning weight scheduling operations.

mean feature A feature having values that are averages among measurement values or other feature values.

mental state value Counterpart to a rapid learning system feature value; within the context of a mental process model.

monitoring Real-time identification of unusual measurement values.

multi-layer perceptron A prediction model that links input measurements to output measurements by a series of composite functions. Each function combines its independent variables linearly and then (typically) transforms the result non-linearly.

multiple regression A linear regression procedure involving one dependent variable and multiple independent variables.

multivariate analysis A linear regression procedure involving multiple dependent variables.

network capacity The number of features that may used within a neuro-computing system.

neuro-computing model A function that includes several inputs and outputs, along with individual processing units, in each of which has inputs are combined linearly according to learned connection weights.

neuron A biological processing unit, including input dendrites, an output axon, and synapses. The term is also used to describe individual processing units within a neural network model.

Notation Conventions

- Greek letters are parameters to be estimated.
- Roman letters are observable measurements.
- Caret overscripts (e.g., $\hat{\nu}$) denote sample-based, (weighted) least-squares parameter estimates.
- cap overscripts (e.g., $\widehat{j}^{\{t'\}}$) denote rapid learning measurement prediction estimates.
- Bold face letters are arrays.
- Entries in subscript parentheses are array dimensions.
- One-dimensional arrays are row vectors.
- Entries in superscript braces (e.g., $\boldsymbol{\nu}^{\{t'\}}$) are time points or record labels.
- Entries in superscript brackets (e.g. $\boldsymbol{\nu}^{[I]}$) are labels.
- **T** superscripts (e.g., $\boldsymbol{e}^{\mathrm{T}}$) denote array transposition.
- **1** denotes an array of 1's.
- **I** denotes the identity matrix.
- $\widetilde{\mathbf{D}}(\boldsymbol{x})$ denotes a diagonal matrix containing the elements of the vector $\boldsymbol{x}$ on its main diagonal.
- Prime superscripts (e.g., t') are counters (dummy variables).
- Plus superscripts (e.g., j^{+}) are maximum values.

off-line analysis The use of a sample to estimate prediction function coefficients; rather than performing real-time learning.

open-loop control Prediction involving the use of dependent variable to influence a process, in a way does not influence future values of independent variables in the prediction function; as opposed to closed-loop control.

optimal estimates Prediction function dependent variable values with minimum distances from observed values in a sample or minimum expected distances from observed values in a target population.

paired associate learning Estimating the prediction function for producing one response from one stimulus during biological learning.

parameter estimate A learned or estimated prediction function coefficient or a related function.

parallel processing Using several computers to compute several input-output functions at the same time.

parity problem Learning how to predict whether the sum of bits in a computer word is odd or even.

PID controller A control device that bases control signals on discrepancies between observed and desired input values, recent changes in such values, and historical averages of such values.

plausibility A number between 0 and 1 used by the rapid learning system to control the impact of an input measurement value on learning.

power feature A feature created by computing powers or products among current measurement values.

prediction Estimating the output value of a prediction function, without actually observing it.

prediction function A mathematical function, with independent variables being measurements that have been observed and dependent variables being measurements that have not yet been observed.

pre-programmed prediction Direct formulation of prediction functions, without using data-based statistical or neuro-computing analysis.

random sample A sample that is created by ensuring that no possible record will be included with a higher likelihood than any other possible record.

random variable A variable that may have different values at two different time points when each other variable being measured has precisely the same value at the two time points.

Rapid Learner™ A trademark, registered by Rapid Clip Neural Systems, Inc, for rapid learning system software owned by them.

rapid learning system An arrangement of computer hardware and/or software that performs adaptive learning, as well as monitoring, forecasting, or control, in real-time.

real-time learning Updating prediction function coefficient values automatically, between measurement record arrival times; as opposed to off-line analysis.

Recent Feature Memory A computer storage medium that contains feature values that were computed from recently observed records.

record An collection of measurement values, the number and order of which is the same within a sample or a process.

recursive updating modifying parameter estimates quickly from record to record, as an alternative to statistical estimation.

redundant measurements/features Measurements/features that may be clustered to form average features without decreasing predictability.

refinement Feature function modification operations performed by the rapid learning system, including redundant feature clustering, unnecessary feature removal, and feature set expansion.

sample A collection of previously gathered records.

separability The capacity for prediction or learning functions to be computed by parallel processors.

simulation Demonstrating rapid learning system operation by processing records from a sample off-line as if they were arriving in real-time.

spatial feature A feature that combines measurements coming from the same physical location.

software Computer programs.

statistic A function of measurement values that typically summarizes a sample or a process in a concise way.

statistical analysis Off-line estimation of prediction function coefficients and other statistics; using data from a random sample.

statistical model An input-output function based on random variables.

standard error Square root of the error variance.

subroutine A computer program that computes one or more frequently used mathematical functions.

Symbol List

- [A] superscript: arithmetic measurement or feature.
- [B] superscript: binary measurement or feature.
- [C] superscript: categorical measurement or feature.
- [D] superscript: dependent variable.
- e: error statistic.
- [E] superscript: error variance.
- $f^{[STEP]}$: forecasting step size.
- $f^{[STEPS]}$: number of forecasting steps.
- [MF] superscript: mean feature.
- [H] superscript: historical feature.
- [I] superscript: independent variable.
- [IN] superscript: independent variable.
- $h^{[STEP]}$: historical step size.
- $h^{[STEPS]}$: number of historical steps.
- j: judgment variable within a mental process model; measurement variable within the rapid learning system.
- l: learning weight.
- m: mental state variable within a mental process model; feature variable within the rapid learning system.
- [M] superscript: multiple correlation coefficient.
- [OUT] superscript: output variable.
- [P] superscript: partial correlation coefficient.
- p: measurement plausibility.
- [S] superscript: spatial feature.
- t: time point.
- v: feature viability.
- w: statistical estimation weight.
- z: (zero mean) standardized score.
- χ: correlation coefficient.
- μ: feature mean.
- ν: feature variance or covariance.
- ω: rapid learning connection weight (feature variance-covariance matrix inverse element).
- ρ: regression coefficient.

telescope A plot of forecast values and their tolerance bands, which typically flare out in the future direction.

training sample A sample that is used for iterative neuro-computing estimation.

Transducer A rapid learning system module that converts input measurements to feature values that are supplied from the Kernel, and converts Kernel output feature values to output measurement values from the system.

univariate prediction The prediction of a measurement value based only prior values of the measurement itself.

unnecessary features Features that do not significantly add to the predictability of other features.

user An information processing specialist who performs monitoring, forecasting, or control.

variance A statistic that indicates how much a measurement varies from time point to time point.

vector An ordered string of measurements.

viability A number between 0 and 1 used by the rapid learning system to control the impact of an input feature value.

weight A coefficient in a prediction function.

weighted statistical estimation Statistical analysis based on giving records different relative impact values, as pre-specified by a statistician.

word A collection of bits used by a computer to store numbers.

Index

A

Abnormal profiles, 9
Abrams, T., 129
Adaptive prediction, 32-33, 48, 249
Additive measurement functions, 175-182
Allen, M.J., 208
Amick, C.M., 59, 129
Analysis of covariance, 124, 205, 249
Analysis of variance, 124, 204, 249
Anderson, T.W., 130, 171, 247
Arbitrage, 17, 249
Arithmetic
  Logic unit, 106
  Monitoring descriptive statistics, 216-218
  Monitoring operations, 216-218
  Variables, 74-76, 123-125, 174, 214-220, 249
Aspirin-Migraine software, 46, 171
Aström, K.J., 28, 58, 247
Axon, 110, 249

B

Backpropagation learning, 45-48, 249
Bayes estimation, 134, 153
Behavioral
  Activity, 116-122
  Psychology, 120
Batch simulation, 67-68

Bickel, P., 59, 130, 171
Binary variables, 74-76, 123-125, 127-128, 174, 193-202, 214, 218-220
Bishop, Y.M.M., 130
Bit, 205, 250
Bivariate prediction, 137-141, 250
Box, J.P., 209

C

Carew, T.J., 129
Categorical variables, 74-76, 123-125, 127, 174, 193-202, 214, 218-220, 250
Chip
  Analog, 52-53
  Digital, 52-53
Client, 66-68
Cluster (see Refinement)
Cognitive
  Process, 120-122
  Psychology, 120
Commodities, 15-17, 82-84, 221-222
Concurrent learning
  And control, 17-21
  And information processing, 1-2
  And model refinement, 22-26
  And monitoring, 5-14
  And profile monitoring, 9
  Operation, 69
Connection weights, 24-26, 101, 115, 244-246, 250
Control, 250
  And forecasting, 85
  Calibration, 21
  Closed-loop, 86, 233-237, 251
  Computer network, 22
  Fuzzy, 21, 167, 251
  Health treatment, 21
  High-speed, 21-22
  Industrial process, 19-21
  Missile tracking, 18-19
  Open-loop, 86, 230-234, 254
  Options, 87-88
  Variables, 86-87
Conjunctive models, 127

Conway, L., 129
Coren, S., 129
Correlation, 238-239, 242-246
Covariance, 123, <u>251</u>
Cox, D.R., 130
Cut-off values, 32

D

Daniel, W.E., 27
Data analysis, 31
Data input specifications, 73-77
Dendrite, 110, <u>250</u>
Deutch, D., 129
Deutch, J.A., 129
Deviance (see Monitoring)
Differential reinforcement learning, 159-161, <u>250</u>
Distance, 217-220, <u>251</u>
Discriminant analysis, 124, 205
Doksum, K., 59, 130, 171
Draper, N.R., 59, 171
Dreaming, 120
Durham, S.D., 28

E

Edwards, R.E. Jr., 27
Error statistics, 217-220
Error variance, 150-153, 214-220, <u>250</u>
Estimation, 18, 135-149 (see also Statistical analysis)
Evolution, 96, 128
Executive (see Rapid learning system)
Exponential family models, 127

F

False alarm, 79
Feature, <u>251</u>
  Compound, 200-207, <u>251</u>
  Functions, 8, 24-25, 51, 64, 101-104, 238-241
  Linear, 93-94
  Non-linear, 93-94

Removal (see Refinement)
Separability, a223-225
Set expansion (see Refinement)
Feedback, 111, 233
Feinberg, S.E., 130
Finn, J.D., 209
Fisher, R.A., 209
Forecasting, 31, 219-226, 251
Commodities, 15-17
Function separability, 223-225
High-speed, 17-18
Steps, 221-222, 226, 251
Telescope, 16, 83-84, 231, 258
Fuzzy (see Control)

G

Garrison, C.Z., 228
General linear models, 123-125, 199, 252
Gleitman, H., 129
Goodman, L.A., 130
Graybill, F.A., 171

H

Hanazawa, T.K., 59
Hardware, 21, 105
Joint access memory, 106-107
Kernel, 106-108
Manager, 108-109
Parallel processing, 51-53, 105-114, 214
Transducer, 108
Hawkins, R.D., 129, 247
Health
Management Monitoring, 13-14
Treatment control, 21
Hidden layer neurons, 43
Hinton, G., 59, 209
Historical
Features, 24, 68, 94-96, 187-190, 207, 252
Steps, 94-96, 188-190
Step size, 94-96, 189-190, 252

Holland, P.W., 130
Hotelling's T^2, 124
Hu, Y., 27, 28, 248
Huntsberger, T.L., 28

I

Identifiability, 50, 127
Image processing, 190-193, 208, 252
Industrial processes, 5-7, 19-21, 235-236
Information processing, 4, 211, 252
Instrument monitoring, 5-7, 10-12
Interaction effects, 204, 252
Interpretability, 34, 36, 41-42, 49-50, 54-55
Iterative neuro-computing estimation, 43-50, 252

J

Jannarone, R.J., 27-28, 59, 129-130, 171, 208, 228, 247
Jeffreys, H., 208
Joint access memory (see Hardware)
Johnson, N.L., 208
Joshi, D., 27
Judgment value, 116-118, 252

K

Kandel, E.R., 129
Kernel, 252 (see also Hardware, Mental process model, Neural network model, and Rapid learning system)
Kosko, B., 28, 58, 171
Kotz, S., 208

L

Lang, K., 59
Laughlin, J.E., 130, 171, 208
Learned Parameter Memory, 105-106, 117-118, 252
Learning, 252
  Impact, 77-79, 159-161, 253
  Specifications, 77-79
  Speed, 34, 40, 49, 51-53

Weight, 77, 101, 104, 111, 123, 149-154, 253
Least squares estimation, 122
Lehmann, E.L., 129-130, 171
Leighton, R., 59
Linear
Prediction, 111, 134, 253 (see also Regression)
Redundancy, 238-239
Lord, F.M., 208

M

Ma, K., 209
Macera, C.A., 228
Mahalanobis distance, 217-220, 253
Manager, 253 (see also Hardware, Mental process model, Neural network model, and Rapid learning system)
Matrix inversion, 51, 122-123, 161
McAvoy, T.J., 28, 58, 247
McClelland, W.L., 58, 129, 209, 248
Mead, C., 129, 208
Mean
Feature, 94-96, 176-182, 252
Statistics, 91-92
Mendel, J.M., 129
Mental process model
Activity, 117-118
Coordination, 118-110
Kernel, 117-118
Manager, 117-120
Mental state value, 118, 253
Transducer, 117-118
Mental state value (see Mental process model)
Mental test theory, 178, 181-182
Meredith, W.M., 209
Miller, R.G., 227
Mind experiments, 154, 171
Missile tracking control, 18-19, 85-86, 232-233
Missing measurements, 74-77, 103-104, 163-167, 171
Model development, 128
Modular structure, 101-106
Monitoring, 253
And Pre-programming, 31

Arithmetic operations, 214-220
Binary and categorical operations, 218-220
Deviance alarms, 13
Deviance statistics, 54, 103, 214-220, 250
Health management, 13-14
High-speed, 14-15
Industrial process, 5-7
Options, 80-83
Statistics, 80-82, 216-218
Strain gauge, 10-12
Structural test, 10-12
Tolerance bands, 13-14
Weld quality, 7-9
Mosteller, F., 59, 171
Multi-layer perceptron, 43, 253
Multivariate analysis (see Regression)

N

Naik, K., 129
Navlani, N., 28
Network capacity (see Rapid learning system Benefits)
Neural network
Biological, 4, 96
Hopfield, 110
Model
Components, 109-111
Kernel, 111, 113-114
Manager, 111, 115
Transducer, 111-113
Neuro-computing, 3, 23, 4, 43-48, 100-109, 254
Neuron, 43-44, 110-115
Neyman, J., 59
Non-linear prediction, 48, 113, 125-126
Non-stationary, 9, 38-39, 42, 50
Notation, 146-147, 254
Novick, M.R., 208
Numerical problems, 161-163, 242

O

Off-line analysis, 7, 8, 17, 20-21, 39, 45, 254

Optimal estimates, 127, 254

P

Pain treatment, 21
Paired-associate learning, 117, 119, 154-158, 254
Palakodety, S.N., 27
Parallel processing (see Hardware)
Parameter estimate, 122, 255
Parity problem, 205-206
Partial correlation, 238-240, 243-245
Pattern recognition, 22-25
Pearl, J., 58, 171
Perception, 121-122
Phoneme recognition, 43-48
PID controller, 21, 39, 31, 235-236, 255
Plausibility, 73-76, 101-102, 149, 177-182, 255
Porac, C., 129
Power feature, 94-96, 183-187, 202-207, 255
Prediction, 255
  Accuracy, 34-35, 41-42, 47-49, 139-141
  And Regression, 134-149
  Function, 136, 138, 142, 147-148, 152, 157, 163, 255
Pre-programmed prediction, 4, 30-37, 255
Prior sample, 149, 153
Product feature, 24, 47, 183-187, 204-207

Q

Qian, D., 27
Quality monitoring, 7-9

R

Random
  Measurement error, 231
  Sample, 38, 255
  Variable, 127, 255
  Walk, 231
Rao, C.R., 228
Rapid Clip Neural Systems, Inc., 5, 6, 8-13, 16, 18, 20, 23-25, 27-28, 34, 41, 44, 48-49, 52, 57, 65, 66-67, 69, 70, 71-74, 76, 78, 80-83, 86-88, 95, 97, 98, 106-

107, 109-110, 112, 114, 116-117, 121, 124, 127, 137, 139-141, 145-146, 151, 155, 157, 160, 162, 171, 191-192, 196, 198, 208, 209, 234, 247
Rapid estimation model, 122-127
Rapid Learner™ software, 255
Dialog box
Control specification, 86-87
Data input specification, 73-76
Feature function specification, 94-96
Forecasting specification, 83-85
Graphical specification, 70-72
Learning specification, 77-79
Monitoring specification, 81-82
Refinement
Specification, 88-91
Statistics, 91-93
Graphical display, 69-73
Graphical user interface, 65-66
Manual, 64, 97
Network capacity, 52-54
Simulator, 67-68
Speed, 51-52
Statistical use, 43
Statistics, 36
User-supplied
Feature function, 94-95
Learning weight file, 77-79
Learning weight function, 77-79
Measurement function, 75-76
Measurements, 75-76
Plausibility values, 75-76
Rapid learning system
Benefits
General, 4
High-speed
Control, 21-22
Forecasting, 17-18
Monitoring, 14-15
Interpretability, 34-36, 41-42, 49-50, 54-55
Learning speed, 34, 41, 49, 51-53
Network capacity, 34-35, 41-42, 53-54
Peak added, 55-57
Prediction accuracy, 34-35, 41-42, 49

Complements, 36-37, 42-43, 50
Examples
Commodities forecasting, 16-17, 82-84, 134
Health management monitoring, 13-14, 134
Health treatment control, 21-22
Industrial process control, 19-21, 235-236
Industrial process monitoring, 5-7, 31
Missile tracking control, 18-19, 85-86, 134, 232-233
Strain gauge monitoring, 11-13, 79-80, 134, 136-141
Visual pattern recognition, 22-25, 134
Voice recognition, 43-48
Weld quality monitoring, 8-9
Executive, 101-102, <u>251</u>
Graphical user interface, 101-102
Kernel, 100-102, 104, 106
Manager, 100-101, 104-106
Feature Function, 101, 104-105
Learning Weight, 101, 104
System, 101
Transducer, 100-102, 106
Real-time operation, 4, 27, 30, 75-76, <u>256</u>
Recent Feature Memory, 101, 105-106, 117-118, <u>256</u>
Recent history features, 103, 187-189, 207
Records, data, 10, 26, 73, 134, <u>256</u>
Recursive estimation, 122, 149-170, <u>256</u>
Redundant
Features, 64, 238-239
Information, 23
Reinforcement, 116-117, 119
Refinement
Clustering, 23, 176-177, 237-239, <u>251</u>
Feature expansion, 24, 88-89, 240-241, <u>251</u>
Feature removal, 24-25, 88-89, 239-240, <u>251</u>, <u>258</u>
Rapid, 25-26, 36
Statistics, 91-93
Technical, 88-89, 242
Regression
Bivariate, 137-140
Formulas, 147-149
General, 38-39, 135-149
Linear, 124, 134, <u>253</u>
Multiple, 124, <u>253</u>

Multivariate, 124, 141-144
Simple, 124
Univariate, 135-137
Weighted, 144-146
Rumelhart, D.E., 58, 129, 209, 248

S

Sample, 38, 40,134, 138, 256
SAS Institute, Inc., 171
SAS™, 38, 46
Sawai, H., 59
Scheffè, H., 130, 209
Separability, 256 (see also Forecasting and Spatial)
Server, 66-67
Shikano, K., 59
Simulation, 64
Smith, H., 59, 171
Sofge, D.A., 28, 58, 247
Software, 21
Spatial
Features, 191-193
Separability, 191-193
Speech recognition, 44-48
Standard deviation, 91-92
Statistical analysis, 37-43, 122-127
Step-wise refinement, 24-25
Stimulus-response activity, 116-117
Subroutine, 84, 257
Symbol list, 257
Synapse, 110
Structural testing, 10-12
Strain gauge monitoring, 10-12, 79-80, 134, 136-141

T

t-test, 124
Takefuji, Y., 130
Tank level control, 19-21, 234-235
Tank level monitoring, 31-32
Tatman, G., 59, 129
Telescope (see Forecasting)

Thinking, 122
Tolerance band, 15, 68, 79-85, 214-220
  Ratios, 91-92
Training, 45, 258
Transducer, 258 (see also Hardware, Mental process model, Neural network model, and Rapid learning system)
Tukey, J.W., 59, 171

U

Univariate prediction, 258 (see also Regression)
Unnecessary features, 258 (see also Refinement Feature removal)
User friendliness, 34, 36, 41-42, 49-50, 55, 258

V

Variance, 139, 148, 165-170
Vector, 147, 258
Viability, 80-82, 101-102, 111, 177-182
Vision, 23
Vital signs, 12

W

Waibel, A., 59
Ward, L.M., 129
Wasserman, P., 129
Weeks, G.E., 27
Weighted
  Statistical estimation, 144-145
  Sum, 31
Weiland, A., 59
Weld monitoring, 7-9
White, D.A., 28, 58, 247
Williams, R.J., 209
Word, 205, 258

Y

Yen, W.M., 208
Yu, K.F., 130, 171

MIX
Papier aus verantwortungsvollen Quellen
Paper from responsible sources
FSC® C105338

If you have any concerns about our products,
you can contact us on
ProductSafety@springernature.com

In case Publisher is established outside the EU,
the EU authorized representative is:
Springer Nature Customer Service Center GmbH
Europaplatz 3, 69115 Heidelberg, Germany

Printed by Libri Plureos GmbH
in Hamburg, Germany